High-*k* Materials in Multi-Gate FET Devices

Science, Technology, and Management Series

Series Editor: J. Paulo Davim, Professor, Department of Mechanical Engineering, University of Aveiro, Portugal

This book series focuses on special volumes from conferences, workshops, and symposiums, as well as volumes on topics of current interested in all aspects of science, technology, and management. The series will discuss topics such as, mathematics, chemistry, physics, materials science, nanosciences, sustainability science, computational sciences, mechanical engineering, industrial engineering, manufacturing engineering, mechatronics engineering, electrical engineering, systems engineering, biomedical engineering, management sciences, economical science, human resource management, social sciences, engineering education, etc. The books will present principles, models techniques, methodologies, and applications of science, technology and management.

Integration of Process Planning and Scheduling
Approaches and Algorithms
Edited by Rakesh Kumar Phanden, Ajai Jain and J. Paulo Davim

Understanding CATIA
A Tutorial Approach
Edited by Kaushik Kumar, Chikesh Ranjan, and J. Paulo Davim

Manufacturing and Industrial Engineering
Theoretical and Advanced Technologies
Edited by Pakaj Agarwal, Lokesh Bajpai, Chandra Pal Singh, Kapil Gupta, and J. Paulo Davim

Multi-Criteria Decision Modelling
Applicational Techniques and Case Studies
Edited by Rahul Sindhwani, Punj Lata Singh, Bhawna Kumar, Varinder Kumar Mittal, and J. Paulo Davim

High-k Materials in Multi-Gate FET Devices
Edited by Shubham Tayal, Parveen Singla, and J. Paulo Davim

Advanced Materials and Manufacturing Processes
Edited by Amar Patnaik, Malay Kumar, Ernst Kozeschnik, Albano Cavaleiro, J. Paulo Davim, and Vikas Kukshal

Computational Technologies in Materials Science
Edited by Shubham Tayal, Parveen Singla, Ashutosh Nandi, & J. Paulo Davim

For more information about this series, please visit: https://www.routledge.com/Science-Technology-and-Management/book-series/CRCSCITECMAN

High-*k* Materials in Multi-Gate FET Devices

Edited by Shubham Tayal,
Parveen Singla, and J. Paulo Davim

CRC Press
Taylor & Francis Group
Boca Raton London New York

CRC Press is an imprint of the
Taylor & Francis Group, an **informa** business

First edition published 2021
by CRC Press
6000 Broken Sound Parkway NW, Suite 300, Boca Raton, FL 33487-2742

and by CRC Press
2 Park Square, Milton Park, Abingdon, Oxon, OX14 4RN

© 2022 Taylor & Francis Group, LLC

CRC Press is an imprint of Taylor & Francis Group, LLC

ISBN: 978-0-367-63968-6 (hbk)
ISBN: 978-0-367-63969-3 (pbk)
ISBN: 978-1-003-12158-9 (ebk)

DOI: 10.1201/9781003121589

Typeset in Times
by SPi Technologies India Pvt Ltd (Straive)

Contents

Preface

The successful investigation of insulated-gate field-effect transistor (FET) device by J. Atalla and D. Kahng at Bell Laboratory in 1959 proved to be the start of a new era in the semiconductor industry. Not long after the progress of complementary metal-oxide semiconductor (CMOS) technology has geared up, the semiconductor market has been taken over by metal oxide semiconductor (MOS) devices. Today's generation of electronic gadgets are so tiny, flexible, and economical that numerous functions can be performed by using a handheld electronic system. The demand for smaller and faster transistors has been fast-tracked by the aggressive scaling of their dimensions. Such a trend has brought about a comprehensive refurbishment of the conventional planar transistors (CPTs) where the single-gate topology has moved to present multi-gate field-effect transistors. Furthermore, new materials such as a high-k dielectric, strain silicon, and metal gates are also a part of present multi-gate devices to enhance their performance. This book deals with the application of high-k dielectrics materials in multi-gate FETs. This book is intended for researchers in the field of semiconductor device modeling and undergraduate/postgraduate engineering courses in the fields of electrical and electronics. Beginning with the motivation behind the multi-gate FETs and high-k dielectrics, the book also covers the various way of utilizing these high-k dielectrics in multi-gate FETs for enhancing their performance at the device as well as circuit levels.

This book has been organized into ten chapters. Chapter 1 presents the motivation behind the multi-gate FETs and current and future trends in transistor technologies. Chapter 2 gives a brief description of the fabrication of high-k dielectrics along with their properties, challenges, and applications to use them with FET devices. Chapter 3 covers the influence of high-k dielectric stacked along with gate engineering on multi-gate FET devices that include Double-Gate (DG) MOSFET, Double Metal DG MOSFET, Triple Metal DG MOSFET, GAA MOSFET, Double Metal GAA MOSFET, and Triple Metal GAA MOSFET. Chapter 4 details the impact of charge trapping in the oxide region or oxide semiconductor interface for further improvement of high-k dielectric-based FET devices. Chapter 5 provides a comprehensive analysis of the impact of high-k dielectrics utilized in the gate-oxide and the gate-sidewall spacers on the GIDL of emerging multi-gate FET architectures. Chapter 6 deals with the application of high-k dielectric materials in various novel FET architectures like Tunnel FETs, Junctionless FETs, Silicon Nanowire transistors, and Carbon Nanotube transistors. Chapter 7 is devoted to high-k gate dielectric-based DG-MOSFET and explains its application in terms of device performance. Chapter 8 gives insights into the usage of high-k dielectric in FET devices for detection of biomolecules, and Chapter 9 deals with the usage of high-k dielectrics for improving the performance of junctionless FET-based SRAM cell. Finally, Chapter 10 covers high-k dielectric-based advanced FETs for lower technology nodes. We are grateful to the contributors of each chapter from renowned institutes and industries, as well as editorial and production teams for their unconditional support in the process of publishing this book.

Editors

Dr. Shubham Tayal is an Assistant Professor in the Department of Electronics and Communication Engineering at SR University, Warangal, Telangana, India. He has more than six years of academic/research experience of teaching at the undergraduate and postgraduate levels. He has received his PhD in Microelectronics & VLSI Design from the National Institute of Technology, Kurukshetra, MTech (VLSI Design) from YMCA University of Science and Technology, Faridabad, and BTech (Electronics and Communication Engineering) from MDU, Rohtak. He has published more than 29 research papers in various international journals and conferences of repute, and many more papers are currently under review. He is on the editorial and reviewer panels of many SCI/SCOPUS indexed international journals and conferences. Currently, he is the editor or coeditor for six books from CRC Press. He acted as the keynote speaker and delivered professional talks on various forums. He is a member of various professional bodies including, among others, IEEE and IRED. He is on the advisory panel of many international conferences. He is a recipient of the Green ThinkerZ International Distinguished Young Researcher Award 2020. His research interests include simulation and modeling of multi-gate semiconductor devices, device-circuit co-design in digital/analog domain, machine learning, and IoT.

Dr. Parveen Singla is Professor in Electronics & Communication Engineering Department of Chandigarh Engineering College–Chandigarh Group of Colleges, Landran, Mohali, Punjab. He received his Bachelor of Engineering in Electronics & Communication Engineering with honors from Maharishi Dayanand University, Rohtak; Master in Technology in Electronics & Communication Engineering with honors from Kurukshetra University, Kurukshetra; and PhD in Communication Systems from IKG Punjab Technical University, Jalandhar, India. He has 17 years of experience in the field of teaching and research. He has published more than 35 papers in various reputed journals as well as national and international conferences. He also organized more than 30 technical events for the students in order to enhance their technical skills and received Best International Technical Event Organiser Award. He is the guest editor of various reputed journals. His interest area includes drone technology, wireless networks, smart antenna and soft computing.

 **Dr. J. Paulo Davim** is a Full Professor at the University of Aveiro, Portugal. He is also distinguished as honorary professor at several universities, colleges, and institutes in China, India, and Spain. He received his PhD in Mechanical Engineering in 1997, MSc in Mechanical Engineering (materials and manufacturing processes) in 1991, Mechanical Engineering degree (5 years) in 1986, from the University of Porto (FEUP), the Aggregate title (Full Habilitation) from the University of Coimbra in 2005, and the DSc (Higher Doctorate) from London Metropolitan University in 2013. He is Senior Chartered Engineer with the Portuguese Institution of Engineers with an MBA and Specialist titles in Engineering and Industrial Management as well as in Metrology. He is also Eur Ing by FEANI-Brussels and Fellow (FIET) of IET-London. He has more than 30 years of teaching and research experience in manufacturing, materials, mechanical and industrial engineering, with special emphasis in machining and tribology. He has also interest in management, engineering education and higher education for sustainability. He has guided large numbers of postdoc, PhD, and Master's students as well as coordinated and participated in several financed research projects. He has received several scientific awards and honors. He has worked as evaluator of projects for ERC-European Research Council and other international research agencies as well as examiner of PhD theses for many universities in different countries. He is the editor-in-chief of several international journals, guest editor of journals, books editor, book series editor, and member of scientific advisory boards for many international journals and conferences. Presently, he is an editorial board member of 30 international journals and acts as reviewer for more than 100 prestigious Web of Science journals. In addition, he has also published (as editor and coeditor) more than 200 books and (as author and coauthor) more than 15 books, 100 book chapters, and 500 articles in journals and conferences (more than 280 articles in journals indexed in Web of Science core collection/h-index 59+/11500+ citations, SCOPUS/h-index 63+/14000+ citations, Google Scholar/h-index 81+/23000+ citations). He was listed in World's Top 2% Scientists by a Stanford University study.

Contributors

Varshini K. Amirtha
National Institute of Technology
Tiruchirappalli, India

Manoj Angara
Department of Electronics and
    Communication Engineering
Koneru Lakshmaiah Education
    Foundation
Vaddeswaram, India

T. S. Arun Samuel
Department of Electronics and
    Communication Engineering
National Engineering College
Kovilpatti, India

T. S. Arulananth
Department of Electronics and
    Communication Engineering
MLR Institute of Technology
Hyderabad, India

Sandip Bhattacharya
HiSim Research Center
Hiroshima University
Hiroshima, Japan

Krutideepa Bhol
School of Electronics Engineering
VIT-AP University
Amaravati, India

Vibhu Goyal
Department of ECE
Government Engineering College
Bharatpur, Rajasthan, India

Ravi Gupta
Department of ECE
Government Engineering College
Bharatpur, Rajasthan, India

Sunil Jadav
Department of Electronics Engineering
J.C. Bose UST, YMCA
Faridabad, India

Amit Kumar Jain
Department of Engineering
Institute for Manufacturing
University of Cambridge
Cambridge, UK

Biswajit Jena
Department of Electronics and
    Communication Engineering
Koneru Lakshmaiah Education
    Foundation
Vaddeswaram, India

Annada Shankar Lenka
National Institute of Technology
Rourkela, India

Shweta Meena
Department of ECE
National Institute of Technology
Kurukshetra, India

Umakanta Nanda
School of Electronics Engineering
VIT-AP University
Amaravati, India

S. V. S. Prasad
Department of Electronics and
    Communication Engineering
MLR Institute of Technology
Hyderabad, India

K. Srinivas Rao
Department of Computer Engineering
MLR Institute of Technology
Hyderabad, India

Shubham Sahay
Indian Institute of Technology
Kanpur, India

Prasanna Kumar Sahu
National Institute of Technology
Rourkela, India

Gaurav Saini
Department of Electronics &
    Communication Engineering
National Institute of Technology
Kurukshetra, India

Ayodeji Olalekan Salau
Department of Electrical/Electronics
    and Computer Engineering
Afe Babalola University
Ado-Ekiti, Nigeria

Trailokya Nath Sasamal
Department of Electronics and
    Communication Engineering
National Institute of Technology
Kurukshetra, India

Shubham Tayal
Department of ECE
SR University
Warangal, India

C. Usha
Department of Electronics and
    Communication
Dayananda Sagar College of
    Engineering
Bengaluru, India

P. Vimala
Department of Electronics and
    Communication
Dayananda Sagar College of
    Engineering
Bengaluru, India

Nisha Yadav
Department of Electronics Engineering
J.C. Bose UST, YMCA
Faridabad, India

1 Introduction to Multi-Gate FET Devices

*T. S. Arulananth, S. V. S. Prasad, and
K. Srinivas Rao*

MLR Institute of Technology, Hyderabad, India

CONTENTS

The development of multi-gate transistors follows several strategies applied by MOS semiconductor manufacturers to create much smaller microprocessors and storage cubicles, colloquially based on Moore's law. Improvement efforts in the field of multi-gate transistors had been reported by the electrotechnical laboratories such as Toshiba, Hitachi, IBM, Grenoble INP, TSMC, Intel, AMD, UC Berkeley, Infineon

DOI: 10.1201/9781003121589-1

Technologies, Samsung Electronics, KAIST, Loose Scale Semiconductor, and others. Other complementary techniques for tool scaling encompass channel pressure engineering, silicon-on-insulator-based total technologies, and excessive-*k*/metallic gate substances. Multi-gate MOSFET structures can acquire greater electrostatic reliability than the traditional planar MOSFETs and, for this reason, offer a path to smaller V_{DD}, lessen V_{TH} variability, and increase transistor scaling.

A GAAFET is abbreviated as a Gate-All-Around (GAA) FET and also as a Surrounding-Gate Transistor (SGT). It is comparable in concept to a Fin-FET except that the gate fabric encompasses the channel vicinity on all the aspects. Relying on layout FET devices can use two or four useful gates. SOI-type multi-gate FETs is much simpler to construct than their bulk complements. Bulk substrate devices when compared with thin substrate device, have more resistance, lower temperature coefficient and stability.

1.1   INTRODUCTION

Since the 1950s, semiconductor electronic device development has increased drastically. Today's electronic market covers the globe and offers potential for improved sophistication of everyday life to every person in the world. Every new invention upgrades or replaces the existing device in the market. The semiconductors provide wide flexibility when it comes to upgrading the modern electronic world. Diodes and transistors have eliminated the need for vacuum tubes. Transistor is known as a linear semiconductor device and is used to control the current that uses less power. Generally, transistors may be grouped into two major categories: bipolar transistors and junction field effect transistors. Bipolar transistors are mainly used to control current ranging from few µAmps to Amps. Field effect transistors (FETs) are electronic semiconductor devices utilizing a lesser voltage to control the current [4, 5]. Among the latest developments are junction field effect transistors (JFETs) and metal oxide semiconductor field effect transistors (MOSFETs) – the insulated gate variety transistors. With continued technological advancements and upgrades, there are many semiconductor devices occupying the electronic markets. Device size, performance, power requirements, package density, and cost are the major selective parameters of the semiconductor devices. Hence, many multi-gate devices introduced to meet the above-mentioned parameters are using self-governing gate electrodes known as multi-impartial gate field effect transistor (MIGFET) [2, 3]. Commonly used multi-gate devices are the FinFET (Fin field effect transistor) and GAAFET (gate-all-around field effect transistor) three dimensional (3D) transistors. In this chapter we discuss the material used for constructing the devices and the performance of the multi-gate FET devices. A multi-gate device (MuGFET) alludes to a MOSFET that joins multiple gates into solitary devices. Thus, the multiple gates might be constrained using solitary gate anode, wherein the various gate surfaces act electrically as a solitary gate. Multi-gate semiconductor devices are one of the few systems being created by CMOS semiconductor producers to make an advanced chip and memory units, conversationally alluding to Moore's law [1]. Categories of multi-gate devices are presented in the following sections.

1.2 DOUBLE-GATE DEVICES

The semiconductor devices that contain two gates are known as double-gate devices. The properties of double-gate devices are [2]:

1. The two gate terminals are associated together.
2. Established electric field lines between source and channel, under the device end with base gate terminal, and cannot subsequently arrive at the channel locale.
3. The field lines spread over the silicon itself can overstep the channel locale and debase small channel attributes. This infringement should be diminished through lessening of the silicon film depth.
4. Originally manufactured double-gate SOI MOSFET devices are completely depleted with the thin channel semiconductor (DELTA, 1989). Figure 1.1 shows the internal structure of DELTA MOSFETs.
5. The FinFET structure is like DELTA, aside from the existence of a dielectric layer referred to as the hard veil edge. This hard veil is utilized to predict the development of dependent reversal channel at the top corner of the devices. Figure 1.2 shows the basic structure of FinFET.
6. Other double-gate MOSFET include Gate-All-Around Device (GAA), Silicon on Noting (SON) MOSFET, and multi-edge XMOS.
7. GAA is a planar MOSFET with the gate cathode folded over the channel district.
8. The MIGFET is a dual gate device wherein the two gate cathodes are not associated together and can, along these lines, be one-sided with various

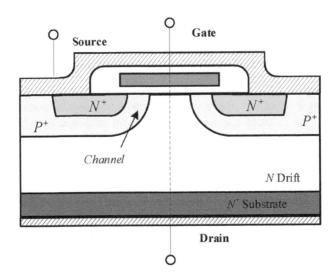

FIGURE 1.1   DELTA MOSFET structure.

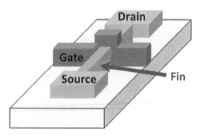

FIGURE 1.2   Basic structure of FinFET.

possibilities. The principle highlights of MIGFET devices are to limit the voltage of one of the gates, which can be regulated by the predisposition given to another gate.

1.3   TRIPLE-GATE MOSFET DEVICES

1. Triple-gate MOSFETs have a thin-film, limited silicon separation between gate terminals.
2. These include the significant wire SOI MOSFET and the tri-gate MOSFET.
3. The introduced electrostatic effect of triple-gate MOSFETs can be enhanced by broadening the sidewall segments of the gate cathode to some profundity in the covered oxide and underneath the channel area π-gate devices and Ω-gate device from an electrostatic perspective; the π-gate devices and Ω-gate MOSFETs have a sustainable number of gates, which may be three or four.
4. Utilization of stressed silicon and metallic gate or potentially dielectric (high-*k*) will improve the flow energy of the devices [6, 7].

1.4   FINFET CHARACTERISTICS AND MODELING

FinFET has got its name because of the field effect transistor (FET) with a structure that appears as a set of fins. It consists of a conducting region, mainly surrounded with the thin "fin" structure and is built on silicon insulator, by which the name "FinFET" has come to be. The device channel for effective conduction is calculated with the fin thickness. Figure 1.3 shows the general structure of FinFET.

The FinFET is a nonplanar device as shown in Figure 1.4; double-gate transistor, which is either a bulk silicon-on-insulator (SOI) or silicon wafers [8–11]. This is based on a single-gate transistor design. There are two different types of FinFET: bulk FinFET and SOI FinFET. The critical characteristic of the FinFET is that it has a conducting channel that is bounded by a thin silicon fin [13]. This mainly forms the body of the device. These fins are nothing but the channel between the source and the drain. The gate terminal is bounded around the channel [12]. This allows the formation of several gate electrodes to reduce the leakage current and to improve the drive current [14].

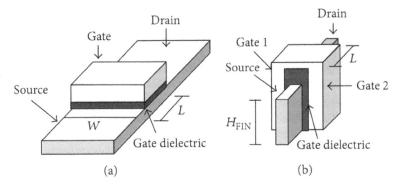

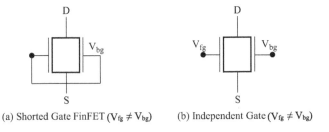

FIGURE 1.3   Difference between MOSFET and FinFET.

FIGURE 1.4   FinFET models.

The FinFET works the same as a conventional MOSFET. It operates in two modes: enhancement and depletion. The working characteristics are identical in both modes; the only difference is that in the enhancement mode, if no voltage is given to the gate terminal, it does not conduct, whereas in the depletion mode, if the voltage is applied to the gate, it does not conduct [16]. In the enhancement mode, when the voltage is applied to the gate terminal, a parallel plate capacitor is formed. The gate is made up of the oxide layer. The surface below the oxide layer is located between the source and the drain. When a small amount of positive voltage is applied to the gate concerning the source, a depletion region is formed. This region is reversed to n-type by the applied positive voltage. Then a region is formed at the interface between Si and SiO_2 [15]. This applied positive voltage attracts the electrons from the source terminal to the drain terminal. In this way, an electron reach channel is formed. The flow of current starts by applying a voltage between the source and the drain. This flow of current is dependent on the voltage applied to the gate [18].

The main advantage of FinFET is that it consumes much less power, which allows higher integration levels [17, 18]. It operates at the lower voltages, as the threshold voltage given is smaller. The FinFET with 20nm technology provide greater switching time and current density. The static leakage current has been reduced up to 90% compared to that of traditional methods, and its operational speed has been increased up to 30% compared to non-FinFET devices.

1.5   SURROUNDING-GATE SOI MOSFETS

This is the type of MOSFET in which silicon or germanium layers form an insulator layer, possibly by a buried oxide layer on the semiconductor substrate, shown in Figure 1.5 [19]. Properties of SOI MOSFETs are listed below:

    i. Surrounding-gate MOSFET structure hypothetically offers the most ideal regulator of the channel locale through the gate and consequently the most ideal electrostatic uprightness.

    ii. Such devices incorporate a CYNTHIA device (roundabout segment devices) and a column encompassing-gate MOSFET (square-segment device).

    iii. SOI MOSFET devices have less than 5 nm gate length and width of 3 nm, which have demonstrated increased usefulness.

1.6   ADDITIONAL MULTI-GATE FETS

Figure 1.6 shows the various forms of multi-gate MOSFETs. Properties and significance of multi-gate MOSFETs are:

    i. The transformed t-channel FET (ITFET) joins a thin-film planar SOI device with a tri-gate semiconductor. It includes planar flat directs and upright diverts in a solitary device.

    ii. Transformed "T" gate assembly has a few preferences: the enormous base helps the balances from dropping over during handling; it additionally considers semiconductor activity in the space between the edges. These extra channels may raise the flow of current in the device.

    iii. The sides of the device switch on first, quickly followed by the outside of the planar areas and the vertical channel.

    iv. Such devices have around seven corner components, and they are establishing more current to each ITFET devices. It can yield significantly more current than a planar device of comparable region.

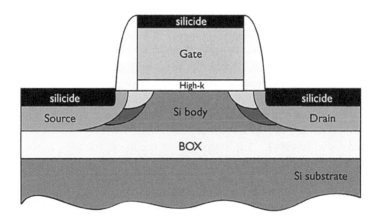

FIGURE 1.5   Structure of SOI MOSFETs devices.

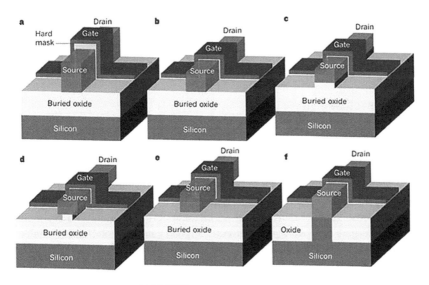

FIGURE 1.6 Multi-gate MOSFET's structures.

v. The Bulk FinFET is a special type of FinFET device which is based on mass silicon instead of a SOI wafer. Fins are carved on a mass silicon wafer and managed utilizing an oxidation step. Devices with balance width down to 10 nm have appeared to have great punch-through insusceptibility down to the sub-20 nm system.

vi. The multi-channel FET (McFET) is an altered mass FinFET. In this device the channel is carved in the focal socket of the balance [20].

1.7 MATERIALS FOR MULTI-GATE DEVICES

The characteristics/performance parameters vary from device to device, mainly based on materials used for constructing a particular device [7]. Major and considerable parameters of the materials are dielectric constant (K), band gap (E_g) in energy band diagram and melting temperature, etc. Some of the material characteristics are discussed in Table 1.1.

TABLE 1.1
Important Semiconductor Material and Its Characteristics

Materials	Dielectric constant (κ)	Bandgap (Eg)	Melting point ($\varnothing_m$)
Silicon dioxide (SiO_2)	3.9	8.9 eV	1,710°C
Hafnium oxide (HfO_2)	25	5.78 eV	2,758°C
Aluminium dioxide (Al_2O_3)	9	8.8 eV	2,072°C

1.7.1 REQUIREMENTS AND CHARACTERISTICS OF MULTI-GATE IN DEVICES

So as to be viable, a device and its characteristics with more than one gate need to meet different prerequisites delineated below and also presented in Figure 1.7.

- Integration with single gate cathode and different gate anode structures
- Alignment of the multiple gate
- Minimal cover of the multiple gates
- Good entryway dielectric nature of the multiple gates

These necessities are normal for both single-gate anode and independent gate terminals variants of various gate devices. The vertical multiple independent gate FET (MIGFET) [2] has come nearest to meeting all the necessities of such device t structures, and its presentation qualities and applications will be examined in detail. Optimal multiple-gate devices require perfectly aligned gate and good-quality dielectrics on both sides of the channel [9].

1.8   NANO-SHEET GAA

Nano-sheets are rising as an enterprise consensus for 5 nm, in accordance to IBM. These devices begin with alternating layers of silicon and silicon germanium (Si-Ge), patterned into pillars. Creating the preliminary Si/Si-Ge hetero-structure is straight-forward, and pillar patterning is comparable to fin fabrication. The subsequent quite a few steps are unique to nano-sheet transistors, though. This spacer defines the gate width. Then, as soon as the internal spacers are in place, a channel launch etch gets rid of the Si-Ge. Si-Ge layers must be as low as possible to decrease lattice distortion and different defects in the germanium content material. Etch selectivity will increase based on the Ge content, though, and erosion of the silicon layers in the course of both the internal spacer indentation and the channel launch etch will have an effect on channel thickness and, consequently, threshold voltage [21].

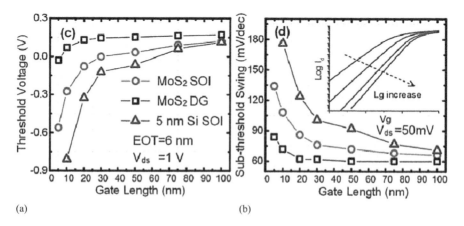

FIGURE 1.7   Characteristics of FET devices: (a) Gate length vs threshold voltage and (b) Gate length vs sub-threshold swing.

1.8.1 Nanowire GAA

In practice, few substances are taking a shot at nanowire gate-all-around FET. For instance, IBM as of late portrayed an gate all-around silicon nanowire FET. This accomplished a nanowire pitch of 30 nm and a scaled gate pitch of 60 nm. The nanowires are framed and suspended on a level plane on the arrival cushions. At that point, vertical entryways are designed over the suspended nanowires. In doing so, numerous gates are framed over a typical suspended locale. A spacer is shaped. At that point, the silicon nanowires cut the need of external gate entry of the device, as indicated by IBM [14].

Utilizing silicon/silicon-germanium super cross-section epitaxial and an in situ doping measure for stacked wires, scientists have built up a stacked, four-wire gate all-around FET. The gate length for the device is 10 nm. Both the channel width and the stature are 10 nm each, in light of an electrostatic scale length of 3.3 nm.

"Limit voltage doping (plans) for stacked wires is far not quite the same as for traditional methodology, particularly when numerous layers of semiconductors are incorporated on a similar substrate," as per the reference [30]. "Leaving the channel un-doped has a bit of leeway in versatility and is required to soothe the issue of arbitrary dopant change, yet it doesn't address the issue for multi-V_t configuration being regularly utilized in SoC (System on Chip) applications. Rather, extraordinary gate work capacities (or gate materials) will be required for various Vt's, and consequently, such un doped approach would be considerably more muddled." Analysts have executed an alternate methodology. "During epitaxial measure, the in-situ doped channel is executed for every one of the stacked wires," as indicated by scientists. "Doped stacked-GAA MOSFETs give adaptable choices to V_t alteration" [18, 22].

1.8.2 SOI Multi-Gate MOSFET Designs

A fin field effect transistor (FinFET) is a type of nonplanar or 3D transistor used in the sketch of cutting-edge processors. As in earlier, planar designs, it is constructed on an SOI (silicon on insulator) substrate [23]. However, FinFET designs additionally use a conducting channel that rises above the stage of the insulator, developing a thin, fin-shaped silicon shape referred to as a gate electrode. The fin-shaped electrode lets in more than one gate to function on a single transistor [11, 24].

1.9 GATE-ALL-AROUND (GAA) NANOWIRE (NW) MOSFETS

So as to build current drive and better control short-channel impacts (SCE), SOI MOSFETs have advanced from planar single-gate to multi-entryways, including twofold, tri-gate FinFET, and gate for what it's worth (GAA) NW FETs in Figure 1.8 [25]. The multi-gate structure gives a superior electrostatic uprightness that portrays how well the gate controls the channel area. The electrostatic honesty is identified with a "characteristic length" λ that can be gotten from Poisson's condition. The normal length speaks to the length of the channel locale that is constrained by the channel. The SCE is stifled if the gate length is greater than the five and up to multiple times of λ. λ diminishes by expanding the quantity of gates [15].

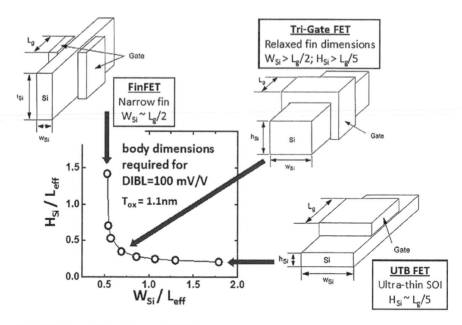

FIGURE 1.8   SOI multi-gate MOSFET designs.

1.10   GAA NANOWIRE (NW) TRANSISTORS

GAA nanowire (GAA-NW) semiconductors are especially reasonable plans for smothering SCE. The overall gate ensures ideal electrostatics, as appeared in Table 1.2, and the one-dimensional quantum repression impacts lessen the transporter dispersing. One can be downsized further by lessening t_{Si} and t_{ox}, as well as utilizing high-*k* dielectric materials to increment e_{ox}. Slim nanowires (<10 nm) with high-*k* as the dielectric material are better than GAA NW FETs [8,30]. We use e-pillar lithography to manufacture nanowires with a distance across of <20 nm, as appeared in Figure 1.9 of a SEM picture for detached 20-nm stressed Si NWs and NWs with a HfO$_2$/TiN gate all around structure. More slender NW can be accomplished by oxidation and carving.

TABLE 1.2
Calculating λ for Various Devices

Single gate

$$\lambda = \sqrt{\frac{\varepsilon_{Si}}{\varepsilon_{ox}} t_{Si} t_{ox}}$$

Double gate

$$\lambda = \sqrt{\frac{\varepsilon_{Si}}{2\varepsilon_{ox}} t_{Si} t_{ox}}$$

GAA structure

$$\lambda = \sqrt{\frac{2\varepsilon_{Si} t_{Si}^2 \ln\left(1 + \dfrac{2t_{ox}}{t_{Si}}\right) + \varepsilon_{ox} t_{Si}^2}{16\varepsilon_{ox}}}$$

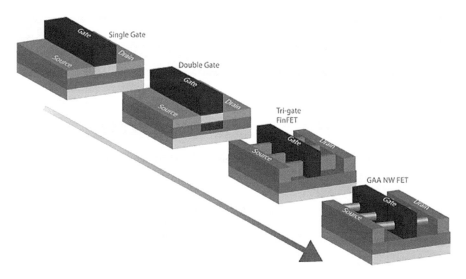

FIGURE 1.9   Evolution of transistor from single-gate SOI FET to GAA NW FET.

1.11   DOUBLE-GATE VERSUS TRI-GATE FET

The double-gate FET shown in Figure 1.10 no longer requires a fantastically selective gate etch, due to the protecting dielectric challenging mask. Additional gate fringing capacitance is much less of a trouble for the tri-gate FET, because the pinnacle fin floor contributes to present-day conduction in the ON state [1, 28]. Power density and variability now restrict traditional bulk MOSFET scaling. Multi-gate MOSFET buildings can acquire greater electrostatic integrity than the traditional planar bulk MOSFET shape and subsequently provide a pathway to decrease V_{DD}, limit V_{TH} variability, and lengthen transistor scaling [7, 29].

1.12   MIGFET AND FINFET PROCESS TECHNOLOGY

The vertical double-gate system with single-gate electrode has emerged as normally recognized as the FinFET which shown in Figures 1.11 and 1.12 shows the overhanging spacer on the silicon fin [5, 6]. The FinFET presents top-notch quick channel manipulation and device traits for digital applications. It is ideal that a few unbiased gate units be built in with these FinFET devices.

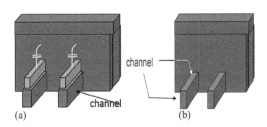

FIGURE 1.10   (a) Double-gate and (b) tri-gate FET.

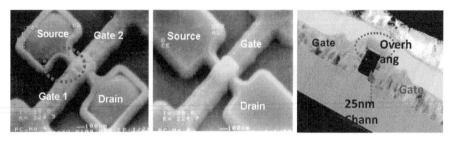

FIGURE 1.11   The MIGFET (a) has independent gates on both sides of thin silicon channels. (b) The MIGFET is easily integrated with finFET.

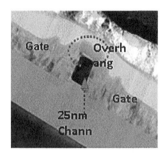

FIGURE 1.12   Overhanging spacer on the silicon fins.

The fins MIGFET devices make such integration viable [12]. This is a maskless system, and the FinFET device are blanketed for the duration of this process. The MIGFET has more advantage of using it in digital CMOS technology as compared to FinFET [20]. The devices have channel dimensions as small as 30 nm and gates of 80 nm. The channel is left undoped, and hence the devices end up being fully depleted when both gates are biased. The process uses 90-nm SOI process techniques and is completed with a cobalt silicide and copper back-end process [13].

1.13   IMPORTANT PARAMETERS IN FINFET

1.13.1   CURRENT

In multi-gate FET, the total current is basically equal to the whole of all streams gushing under the gate terminal (carriers have the same mobility under all the gates). As we increase the amount of passage, the current gets copied by the assessment of current in single-gate FET (keeping all the portal estimation same). Current in multi-gate FET is calculated by (Number of gate) * (current in single-gateway FET). We can say that current in twofold-passage FET is twice that of the single-gate FET of identical gate length and width. As we increase the number of gates, it will fabricate the current any time it has obstructions, i.e., difficult to produce, corner impacts, etc. For example, a three-sided edge will have a twofold entryway, square equalization will have a triple-passage and pentagonal parity will have a quad-gate around it.

Instead of extending the number of entryways, we go for multi-cutting edges of essential shapes (i.e., rectangular, square). We can get tremendous current if we increase the cutting edges in a device [16].

By growing sharp edges, we will get part of central focuses that shown in Figure 1.14a and b:

- Large current.
- Area of device gets decreased.
- Fabrication is pinch straightforward.

Current in multi-sharp-edge FET = No. of fins * current in single cutting edge of contraption. Figure 1.13a and b shows the top and side views of the FinFET devices.

As fin height increases, the current also increases, but it is a bit difficult to fabricate large-height fin devices, hence the proposed low-height multi-fin construction.

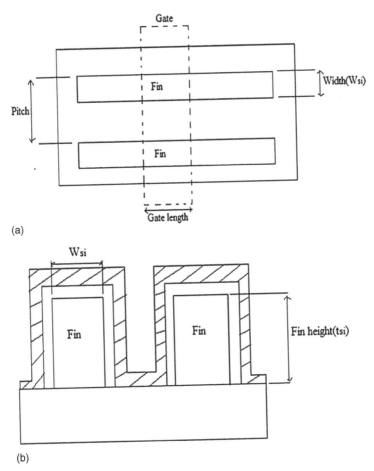

FIGURE 1.13 FinFET: (a) top view and (b) side view.

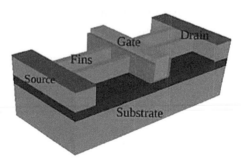

FIGURE 1.14   Two-fin FinFET.

1.13.2 CORNER EFFECTS

When a restriction on Fin shapes exists, corner impact comes into the picture. Corner effects lead to the untimely channel shaping at the edges of the Fin. It causes the leakage current in the device in light of the fact that the edge voltage of this untimely channel is low. Untimely channel can shape at the corner in multi-gate FinFET in light of charge-sharing impact between two gate anodes. We can see that top corners of the FET are constrained by top gate and sidewall gates. So on top of the balance we have two untimely diverts in corner with various threshold voltages, or we can say that multi-V_t untimely directs in corners. These channels lead to leakage flows in FinFET. These corner impacts debase the sub-limit attributes of the device. Preferably sub threshold slope ought to be 60 mV/decade, yet because of this it builds, which is terrible for a device [3]. Figure 1.15 shows the corner effect removal in Fin devices.

1.13.3 SUB-THRESHOLD SLOPE

Short channel effect emerges when the separation between source and channel is limited. This boundary ought to be limited as much as could be expected. Whenever the gate voltage also increases, the channel current also increases. So as little change in gate voltage as possible every decade is required. Sub-threshold slope is given by

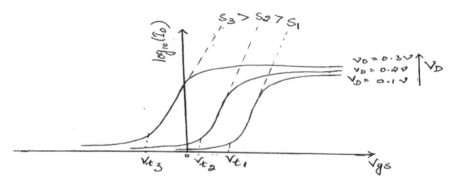

FIGURE 1.15   Sub-threshold slope for the cases S3 > S2 > S1.

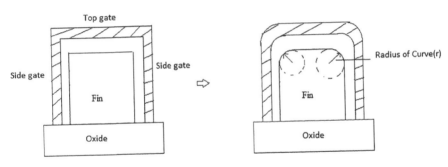

FIGURE 1.16 Corner effect removal in FIN.

= $dV_g/dlog(Id)$ shown in Figure 1.14. When channel voltage builds, the sub-threshold slant will expand, hence the threshold voltage is getting diminished. If V_t is decreased, the leakage current gets increased.

1.14 CONCLUSION

In this chapter, we have discussed in detail about the material for multi-gate devices and other necessities of the multi-gate devices. We prove the profitability of improvement offered by double-gate units that have a single-gate electrode on more than one aspect, which can be used to similarly scale CMOS digital technological know-how and the MIGFETs that have a couple of unbiased gates. The MIGFET that has two impartial gates self-aligned on each aspect of a skinny silicon channel and its utility in digital and analog mixed-signal functions have been explored. The MIGFET-T with three impartial gates can be used in single transistor reminiscence devices. All these units use nicely understood substances and interfaces and can be built in with every different the usage of exiting technique technologies. This allows these combined sign devices to be scaled and built in with the digital CMOS devices even as they are continually scaled. New circuits that use MIGFET have been validated, such as dynamic V_t modulation, RF mixers, and single transistor reminiscence elements. These SOI-based devices need to drastically allow new merchandise and functions when mixed with the superior overall performance of ultra-thin physique double-gated devices. The FinFET-based circuits give the required output by reducing the delay and power consumption at ultra-low supply voltages.

REFERENCES

[1] Mathew L et al., *Proceedings of the IEEE International SOI Conference*, 2004, p. 187.
[2] Hisamoto D et al., *Electron Devices Meeting*, 2001. Technical Digest, p. 429.
[3] Hisamoto D et al., *Electron Devices Meeting*, 1989. Technical Digest, p. 833.
[4] Mathew L et al., *IEEE 2002 Silicon Nano-Electronics*. Work, p. 5.
[5] Mathew L et al., Multiple Independent Gate Field Effect Transistors – Device, Process, Applications, University of Florida, 2009.
[6] https://semiengineering.com/knowledge_centers/integrated-circuit/transistors/3d/gate-all-around-fet/

[7] Kedzierski J, Ieong M, Kanarsky T et al. (2004) Fabrication of metal gated FinFETs through complete gate silicidation with Ni. *IEEE Trans Electron Devices* 51(12):2115–2120.

[8] Yang J-W, Fossum JG (2005) On the feasibility of nanoscale triple-gate CMOS transistors. *IEEE Trans Electron Devices* 52(6):1159–1164.

[9] http://www.signoffsemi.com/finfet-2/

[10] Taghipour S, Asli RN (2013) Aging comparative analysis of high- Performance FinFET and CMOS flip-flops, *Microelectronics Reliability* 69(2):52–59.

[11] Abdelkader O, El-Din MM, Mostafa H, Abdelhamid H, Fahmy HAH, Ismail Y, Soliman AM (2018) Technology scaling roadmap for FinFET-based FPGA clusters under process variations, *Journal of Circuits, Systems and Computers* 27(4):1850056(1–32).

[12] Knoll L, Schäfer A, Trellenkamp S, Bourdelle KK, Zhao QT, Mantl S (2012) Nanowire and planar UTB SOI Schottky barrier MOSFETs with dopant segregation, *Proceedings of the 13th International Conference on Ultimate Integration on Silicon (ULIS),* pp. 67–70.

[13] Richter SS, Trellenkamp S, Schmidt M, Schäfer A, Bourdelle KK, Zhao QT, Mantl S (2012) Strained silicon nanowire array MOSFETs with high-k/metal gate, *Proceedings of the 13th International Conference on Ultimate Integration on Silicon (ULIS),* pp. 75–78.

[14] Tsu-Jae King Liu, *Introduction to Multi-Gate MOSFETs.* Department of Electrical Engineering and Computer Sciences University of California, Berkeley, CA.

[15] Basker V et al. (2010) A 0.063 µm^2 FinFET SRAM cell demonstration with conventional lithography using a novel integration scheme with aggressively scaled fin and gate pitch, *Symposium on VLSI Technology,* pp. 19–20, IEEE.

[16] Guillorn MA et al. (2011) A 0.021 µm^2 trigate SRAM cell with aggressively scaled gate and contact pitch, *Symposium on VLSI Technology,* pp. 64–65, IEEE.

[17] Kavalieros J et al. (2006) Tri-gate transistor architecture with high-*κ* gate dielectrics, metal gates and strain engineering, *Digest of Technical Papers – Symposium on VLSI Technology,* pp. 50–51, IEEE.

[18] Cao S, Chun JH, Salman AA, Beebe SG, Dutton RW (2011) Gate-controlled field-effect diodes and silicon-controlled rectifier for charged-device model ESD protection in advanced SOI technology, *Microelectronics Reliability* 51: 756–764.

[19] Rauly E, Iñiguez B, Flandre D (2001) Investigation of deep submicron single and double gate SOI MOSFETs in accumulation mode for enhanced performance, *Electrochemical and Solid-State Letters.* https://doi.org/10.1149/1.1347225 4, G28.

[20] Lee C-W, Lederer D, Afzalian A, Yan R, Dehdashti N, Xiong W, Colinge JP (2008) Comparison of contact resistance between accumulation-mode and inversion-mode multigate FETs, *Solid-State Electronics* 52: 1815.

[21] Yan R, Lynch D, Cayron T, Lederer D, Afzalian A, Lee C-W, Dehdashti N, Colinge JP (2008) Room-temperature low-dimensional effects of Pi-gate SOI MOSFETs, *Solid-State Electronics* 52: 1872.

[22] Colinge JP et al. (2010) Nanowire transistors without junctions, *Nature Nanotechnology* 5: 225–229.

[23] Zhang Q, Zhao W, Seabaugh A (2006) Low-subthreshold-swing tunnel transistors, *IEEE Electron Device Letters* 27: 297–300.

[24] Zaman RJ, Mathews K, Xiong W, Banerjee SK (2007) trigate fet device characteristics improvement using a hydrogen anneal process with a novel hard mask approach, *IEEE Electron Device Letters* 28(10):916–918. doi:10.1109/LED.2007.905964.

[25] Minhaj EH, Abdur Razzak M, Islam MM, Adnan MMR (2019) Performance enhancement of multigate FinFETs by using high-k stack oxide, *2019 1st International Conference on Advances in Science, Engineering and Robotics Technology (ICASERT)*, Dhaka, Bangladesh, pp. 1–4. doi:10.1109/ICASERT.2019.8934656.

[26] Colinge J (2014) Multigate transistors: pushing Moore's law to the limit, *2014 International Conference on Simulation of Semiconductor Processes and Devices (SISPAD)*, Yokohama, Japan, pp. 313–316. doi:10.1109/SISPAD.2014.6931626.

[27] Sohn C et al. (2012) Device design guidelines for nanoscale FinFETs in RF/analog applications, *IEEE Electron Device Letters* 33(9):1234–1236. doi:10.1109/LED.2012.2204853.

[28] Kumar N, Chen J, Kar M, Sitaraman SK, Mukhopadhyay S, Kumar S (2019) Multigated carbon nanotube field effect transistors-based physically unclonable functions as security keys, *IEEE Internet of Things Journal* 6(1):325–334. doi:10.1109/JIOT.2018.2838580.

[29] Moran JM, Mahoney GE, DiLorenzo JV (1976) Fabrication of multigate power GaAs FETs using electron lithography, *1976 International Electron Devices Meeting*, Washington, DC, pp. 446–449. doi:10.1109/IEDM.1976.189080.

[30] Wang J, Solomon PM, and Lundstrom M (2004) A general approach for the performance assessment of nanoscale silicon FETs, *IEEE Transactions on Electron Devices* 51(9):1366–1370. doi:10.1109/TED.2004.833962.

2 High-*k* Gate Dielectrics and Metal Gate Stack Technology for Advance Semiconductor Devices
An overview

Vibhu Goyal
Government Engineering College, Rajasthan, India

Shubham Tayal
SR University, Warangal, India

Shweta Meena
National Institute of Technology, Kurukshetra, India

Ravi Gupta
Government Engineering College, Rajasthan, India

Sandip Bhattacharya
Hiroshima University, Hiroshima, Japan

CONTENTS

DOI: 10.1201/9781003121589-2

2.1   INTRODUCTION

There has been an enormous advancement in microelectronics industry in the last three decades with a steadily expanding performance of integrated circuits (ICs). This advancement is conceivable due to downscaling of key element of these ICs, that is, MOSFET – metal oxide semiconductor field effect transistor. The downscaling of these elements enables the amalgamation of a large number of FETs on a chip, which in turn leads to improved speed and increased functionality. Hitherto, the downscaling of these FETs has been accomplished via scaling the thickness of nitride silicon oxide (SiON) or silicon dioxide (SiO_2) gate dielectric. Both these materials are used as gate insulators owing to several important properties. First, both of these materials naturally form a very steady interface with the silicon substrate with negligible interface defects. Second, an amorphous layer of both these materials may be grown thermally on silicon with exceptional uniformity and better control on thickness. Third, these materials show an astounding thermal stability, which is the prime concern while fabrication of advanced MOSFETs. Further, energy band gap of silicon dioxide (SiO_2) is sufficiently high to provide admirable electrical isolation and is truly necessary to lessen the gate tunneling current.

However, incessant miniaturization becomes problematic due to various issues concerning heat dissipation and excessive power consumption in ICs. Undeniably, the leakage current coursing through the FETs with sub-1 nm thickness, which arises from direct tunneling of charge carriers, lies well over the limit as recommended by the International Technology Roadmap for Semiconductors (ITRS) [1]. Therefore, microelectronics industry scouts for alternatives to the modern CMOS technology that can boost device performance.

The different approaches include introduction of low-*k* interconnects, high-*k* dielectrics, high channel mobility materials like germanium, GaAs, etc., and nonplanar modern CMOS device structures like FinFET. Since the downscaling of gate oxide (SiO_2) has already touched on the essential material limits, further downscaling may be achieved only by introduction of new materials with high value of dielectric constant (*k*) [2–7].

This chapter starts with a concise description of issues related to the downscaling of gate oxide that led to the integration of high-*k* dielectric materials. Later, we discuss at length the material science of high-*k* dielectrics employed in manufacturing of transistors. Since electrical performance of transistors is of prime importance, we also review the recent developments in transistor gate stack technology-based devices. Defects that assume a significant part in selection of high-*k* dielectrics for FET applications are also discussed besides recent developments in the devices like TFET.

2.2 DOWNSCALING ISSUES AND HIGH-*k* MATERIALS RELEVANCE IN MICROELECTRONICS INDUSTRY

The conventional complementary metal oxide semiconductor (CMOS) devices have gained popularity owing to their low power consumption. The unceasing advancement in their performance over the past more than four decades made it possible to follow the Moore's law, which states that the number of devices on an IC doubles every 18 months. Consequently, the transistor size shrinks exponentially each year.

SiO$_2$ is considered to be the most commonly used gate dielectric in field effect transistors (FETs). However, due to continuous downscaling, this SiO$_2$ layer is now so reedy that leakage current through the gate oxide starts exceeding 1 A/cm^2 at 1 V owing to very high tunneling of electrons through the gate dielectric, thereby exceeding static power dissipation beyond acceptable constraints [3,8–18]. Further, it becomes even more difficult to meet the reliability requirements, and therefore SiO$_2$ has to be substituted by some other material.

It is a universal fact that MOSFET is a capacitive device wherein drain to source current hinges on gate capacitance (C), which can be articulated as

$$C = \frac{\varepsilon_0 K A}{t_{ox}}$$
(2.1)

or

$$\frac{C}{A} = \frac{\varepsilon_0 K}{t_{ox}}$$
(2.2)

Where ε_0, A, and K & t_{ox} represent the free space permittivity, area, relative dielectric constant, and oxide thickness of dielectric, respectively. It is noted here that the tunnel current decreases exponentially with increasing distance [19].

The solution to the aforesaid problem is to replace SiO$_2$ ($K = 3.9$) with a dielectric having higher permittivity than that of SiO$_2$. This can provide equivalent oxide thickness (EOT) at higher physical thickness [refer Figure 2.1] as can be elucidated from Equation 2.3. This EOT will not only retain the same capacitance but will also lessen the tunneling current. This new gate oxide is frequently referred as high-*k* oxide.

$$EOT = \frac{(3.9)}{K} * t_{ox}$$
(2.3)

Several high-*k* materials (ranging from $k \sim 8$ like Al$_2$O$_3$ to $k \sim 103–105$ like perovskites) have been investigated by the researchers so as to come up with a promising substitute for SiO$_2$. Nevertheless, to find an apt high-*k* dielectric is a foremost task, as it should be thermally stable, must have high resistivity, and should be able to work as a barrier layer so as to prevent leakage current. Also, the high-*k* material should provide an idyllic interface with silicon.

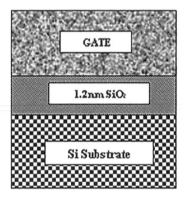 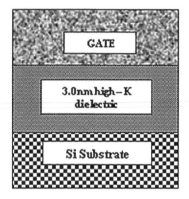

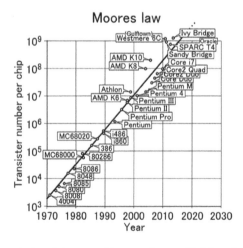

FIGURE 2.1 SiO_2 based gate stack v/s. gate stack of high-*k* material.

FIGURE 2.2   Increase in transistors count in accordance with Moore's law.

2.3   MATERIAL CHEMISTRY AND REQUIREMENTS FOR HIGH-*k* GATE DIELECTRICS

The primary reason why microelectronics industry preferred silicon (Si) technology and not any other semiconductor is SiO_2. Silicon as a semiconductor is an average performer; however, SiO_2 is an excellent insulator in almost every aspect. The main advantage that SiO_2 possess is that it may be formed from silicon simply by the process of thermal oxidation, whereas other semiconductors (i.e., Ge, GaAs, GaN etc.) form a meager interface with their oxide counterparts. Besides this, SiO_2

may be etched and patterned to sub-nanometer level. The only challenge that SiO_2 poses is that tunneling is possible across it when it is very thin. Due to this drawback, Si losses its sheen and is a big impediment in further downscaling of CMOS-based devices. Therefore, device engineers start using alternative high-*k* oxides. We can choose these oxides from a large pool of available elements in the periodic table. However, the necessities of new, thicker gate oxides are manyfold [3,8,11,16–18]:

- It should have large energy band gap with barrier height to silicon substrate as well as to metal gates for reducing the leakage current.
- It must have high value of dielectric constant so that it can be economically viable for a reasonable number of scaling nodes.
- It must be thermodynamically stable on silicon to avoid the creation of a low-*k* SiO_2 interface.
- The high-*k* interface surface density must be low ($\sim 10^{10}$ cm^{-2} eV^{-1}) and must form a good electrical interface with Si.
- It should have better reliability and decent life time.
- It should have high amorphous-to-crystalline transition temperature so that a stable morphology can be maintained after heat treatment.
- Channel carrier mobility should be high enough (~90% of SiO_2/Si system).
- It should have low oxygen diffusion coefficients so that the formation of thick low-*k* interfacial layer can be controlled.
- It must have good kinetic stability and should withstand processing temperature as high as 1000°C for at least 4–5 seconds.

Besides all of the above, the high-*k* dielectric should be adaptable to the current CMOS fabrication techniques and to other materials used in CMOS IC process. In the following subsections, the vital requirements for high-*k* dielectrics are discussed in depth.

2.3.1 ENERGY BAND GAP, BARRIER HEIGHT, AND DIELECTRIC CONSTANT

To date, there is no single dielectric available that can fulfill all the necessities for an ideal gate dielectric. As described above, it is vital to replace SiO_2 with a material having high-*k* value (dielectric constant) as a gate dielectric.

Notations E_V, E_C, E_F, and E_i in Figure 2.3 indicate valence band, conduction band, Fermi level, and intrinsic level, respectively. Figure 2.3 represents metal semiconductor band alignment for n-type semiconductor. The electron affinity and work function for the semiconductor are marked properly (adapted from S. M. Sze's *Physics of Semiconductor Devices*). Without any external bias, the device works in equilibrium mode; i.e., the Fermi levels of both metal and semiconductor are the same. However, with the application of external bias, it behaves in concurrence with the voltage applied with respect to flat band voltage and threshold voltage. When no charge is present at the oxide–semiconductor interface, the work function difference between semiconductor and metal gate is called flat band voltage, whereas threshold voltage is defined as the minimum gate-to-source bias that

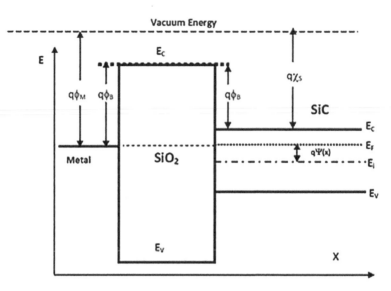

FIGURE 2.3 Band diagram schematic of MOS structure.

is required for creation of a conduction channel. This can be further categorized into three types, namely:

a. Accumulation layer
b. Deletion layer and
c. Inversion layer

In case of an accumulation layer, bias applied at the gate terminal is negative, i.e., it is less than the flat band voltage. The MOSFET will no longer remain in equilibrium. There is a shift in the Fermi level by a magnitude that equals charge of an electron multiplied by applied bias. Thus, there is a rise in metal Fermi level while Fermi level of the semiconductor remains constant.

In depletion layer, applied bias is less than the threshold voltage but greater than the flat band voltage. Since applied bias is positive, there is a fall in the Fermi level of metal while the Fermi level of semiconductor rises. Recombination of the electrons with holes near oxide–semiconductor junction takes place, which creates a depletion region.

In the inversion layer, applied bias is greater than the threshold voltage. The name "inversion layer" derives from the fact that the surface is inverted from p-type to n-type near the oxide–semiconductor junction. Since applied bias is very high, the Fermi level of the metal is further diminished.

2.3.2   LARGE *k* VALUE

The second prime necessity is oxide's *k* value, which must be rather large, around 25–35; practically, it needs to be above 12. However, there is a trade-off between the *k* value and band offset condition. There are several materials with a very high value of *k*, but their band gaps are very small and cannot be considered. Consequently, an oxide with relatively low *k* value can be considered [20].

2.3.3 THERMODYNAMIC STABILITY

The third requirement is the thermodynamic stability of the potential gate oxide insulator on silicon surface. Nowadays, fabrication of devices typically takes place at very high temperatures (>1000°C), and subsequently the dielectric material must retain its solid state during the fabrication process. The oxide, therefore, must have higher heat of formation (per O atom) as compared to SiO_2 [21] and should not react with silicon to form either SiO_2 or a silicide as:

a. $MO_2 + Si = M + SiO_2$
b. $MO_2 + 2Si = MSi + SiO_2$

The SiO_2 layer formed by equation (a) can increase the equivalent oxide thickness and thus negates the effect of new oxide, and hence must be avoided. On the other hand, silicides formed by equation (b) are metallic in nature and may short-circuit the channel. Moreover, it will degrade the electrical properties of high-*k* gate stack structure. Thus, we are left with limited choice of oxides like SrO, BaO, Cao, ZrO_2, HfO_2, Al_2O_3, Y_2O_3, and lanthanides (refer to columns II, III, and IV of the periodic table). Besides this, it is found that group II oxides have low-*k* values and therefore are less useful for this purpose. This leaves us with HfO_2, Al_2O_3, Y_2O_3, and some lanthanides (Lu_2O_3, Gd_2O_3, etc.).

2.3.4 KINETIC STABILITY

The high-*k* material must be adaptable to the existing process conditions and must be kinetically stable, which means the oxide should remain amorphous when annealed for 5 seconds at 1,000°C. Most of the oxides crystallize at below 1,000°C. Unlike SiO_2, which is an exceptional glass former, almost all high-*k* oxides are bad glass formers. The crystallization problem of other oxides may be dodged by alloying the oxides with a glass former like SiO_2 or Al_2O_3, which gives either a silicate or aluminates [6, 8, 11, 13, 17, 20, 22, 23]. These silicates or aluminates retain stability against crystallization up to nearly 1,000°C, though it is found that silicates have rather smaller *k* values. Besides this, it is observed that the addition of nitrogen (N_2) lowers the diffusion rates and raises crystallization temperature [8]. Moreover, nitrogen improves electrical properties as well. Thus HfO_2/Hf silicates satisfy this entire criterion.

2.3.5 HIGH-QUALITY INTERFACE AND DEFECTS

The superb advantage of silicon as a semiconductor is its exceptionally top-notch $Si–SiO_2$ interface (Interface trap density $D_{it} \sim 10^{10}$ $eV^{-1}cm^{-2}$). Top notch implies roughness and absence of deformities to evade scattering carriers. An electrically active defect implies such atomic configuration that can offer ascent to electronic states in the oxide band gap that can trap carriers.

Several reasons are there for which defects are considered undesirable. The prime one is trapping of charge at the defect site that can change the threshold voltage (V_{th}) of the FET – threshold voltage is the voltage at which the transistor turns ON. The second reason is that these trapping charges changes with time, which in turns changes V_{th}, thereby leading to operational instability. Third, these trapped charges scatter the carriers in the channel and thus degrade their mobility. Fourth, defects result in poor reliability, as they are the origin behind the device electrical failure and oxide breakdown.

For high-*k* materials, similar interface quality is expected with Si. However, it is found that most of the high-*k* materials show significant degrees of high interface state density and substantial shift in flat band voltage (ΔV_{FB}) primarily owing to high density of fixed charges [24]. Further, some research works (Luckovsky et al.) find that bonding restraints of high-*k* materials assume a critical part in assurance of high-*k*/Si interface quality. Experimental data show that if the average number of bonds per atom is greater than 3 for a metal oxide, it forms over-constrained high-*k*/Si interface and its interface trap density increases exponentially. Contrary to this, if a metal oxide has coordination number less than 3, it forms under-constrained high-*k*/Si interface that leads to high interface state density but poor device performance.

2.4   OXIDE DEPOSITION

High-*k* layers/materials can be deposited via different deposition techniques as available; each having its own (de)merits. In the following subsection, we have concisely deliberated the most commonly used high-*k* deposition techniques [12] such as metal organic chemical vapor deposition (MOCVD), atomic layer deposition (ALD), and physical vapor deposition (PVD), among others. It should be noted here that contrary to CVD (chemical vapor deposition), PVD offers an advantage of not getting any contamination from the precursor.

2.4.1   MOCVD

High-*k* materials like hafnium oxide (HfO_2) and silicate layers are deposited using MOCVD. As the precursors are liquid in this case, it limits the formation of particles during time of deposition. Generally, the configuration preferred to deposit layers is showerhead configuration, which ensures uniformity across the wafer. The deposition temperature range for MOCVD is 400°C to 600°C. The composition can be varied with the help of variation in gas and pressure flow through the chamber.

2.4.2   ALD

It is a subclass of chemical vapor deposition that depends on the consecutive utilization of gas-phase chemical process. In ALD, we utilize at least two precursors that react with the material surface with each in turn in successive, self-restricting manner. Typically process conditions are temperature in the range of 50°C–500°C and pressure range 0.1–10 mbar.

ALD offers various advantages; a few of them are:

a. Precise control over the thickness at true nanometer scale
b. Highly repeatable and scalable process
c. Pinhole-free films

2.4.3 PVD

The major concern while depositing high-*k* material is the control of SiO_2 interfacial layer. Since there are several challenges in CVD processes like contamination due to hydrogen and carbon, to name a few, an alternative approach to CVD is PVD (sputtering) followed by controlled oxidation (CO). Physical vapor deposition is the process wherein the material goes from a condensed phase to a vapor phase and then back to a thin-film condensed phase. The most commonly used PVD processes are sputtering and thermal/vacuum evaporation. In vacuum evaporation, vaporization of source material is done by heating the material using appropriate methods in vacuum (generally we use a molybdenum boat to keep the target deposited material). The sufficient vacuum required for deposition is achieved with the help of rotary and turbo pumps, whereas in sputtering, which is a plasma-assisted technique, creation of vapor from the source target through bombardment with accelerated gaseous ions (gas generally used is argon) takes place. In both evaporation and sputtering, the resulting vapor phase is subsequently deposited onto the desired substrate via condensation mechanism.

The typical procedure for physical vapor deposition is as follows: (i) sputtering/evaporation of different components to produce a vapor phase; (ii) super-saturation of the vapor phase in an inert atmosphere to promote the condensation of metal nanoparticles; and (iii) consolidation of the nanocomposite by thermal treatment under inert atmosphere.

2.5 METAL GATES AND THEIR WORK FUNCTION REQUIREMENTS

The gate electrode serves the purpose of shifting the surface Fermi level E_F of the channel to the other band edge, to invert the transistor. We know that a PMOS device has an n-doped Si channel and a gate with work function (~5 ev), which shifts its Fermi level (E_F) to its valence band, thereby inverting the channel. Likewise, a NMOS device is comprised of p-doped Si channel and a gate with low work function (~4.0 ev), which can move its surface Fermi level from valence band to conduction band (CB). Since SiO_2 is a perfect insulator, a work function applied on the surface of thin SiO_2 layer gives a work function of the same value at the Si: SiO_2 interface. However, in case of high-*k* dielectrics like HfO_2, it has relevance with Schottky barrier height (SBH). Various research groups [3, 25] have experimentally calculated the SBH for epitaxial interfaces to have in-depth understanding of metal HfO_2. The prime criterion for this calculation is that metal must be lattice matched with HfO_2. Further, it should be noted that the interface can be polar or nonpolar.

2.6 ELECTRICAL BEHAVIOR/ELECTRICAL CHARACTERISTICS OF HIGH-*k*-BASED DEVICES

The usage of high-*K* gate dielectric permits one to radically lessen the leakage current (gate tunneling current) through the MOS devices. Still, numerous challenges need to be overcome that pertains to electrical performance of high-*K*-based devices. Some of the challenges are discussed in depth in the following subsection.

2.6.1 FLAT BAND VOLTAGE AND THRESHOLD VOLTAGE CONTROL

The very major problem encountered in the devices that integrate the high-*k* layers with poly-silicon gates is the control over flat band voltage (V_{FB}) and threshold voltage (V_{th}). Several works have reported the shifts in V_{FB} and V_{th} in reference to ideal SiO_2/poly-Si reference system [12]. The shifts are asymmetrical, being positive for n+-poly-Si and negative for p+-poly-Si. Hobbs et al. [16] projected that interaction of Hf atoms with the poly-Si layer creates a large density of defects in Si band gap that is capable enough to pin the Fermi level. Consequently, this will change the metal work function of the gate electrode, thereby changing the flat band voltage of the device. One more research group, Cartier et al. [10], have noticed similar shifts in flat band voltages both in undoped poly-Si layer and fully silicide gates. It should be noted here that elucidations of shift in V_{FB} are largely based on the eradication of V_{FB} from capacitance–voltage measurements.

2.6.2 MOBILITY DEGRADATION

The prime goal of downscaling is to produce small and ultra-fast devices. Further, it is a well-known fact that a high drain current is required for high-speed devices, which in turn rest on mobility of the carriers. The carrier mobility typifies how fast an electron can move through a metal or semiconductor when driven by an electric field. When an electric field, E, is applied across a piece of material, the carriers (electrons/holes) respond by moving with an average velocity, called drift velocity. The carrier/electron mobility is defined as

$$V_d = \mu E$$

We know mobility is controlled by different mechanisms at different gate fields: that is, at low fields, it is bounded by columbic scattering; at moderate fields it is restricted by acoustic phonon scattering; while at high fields, it is confined by interface roughness scattering. All of these scattering mechanisms have different temperature dependences. Out of these three scattering mechanisms, phonon scattering is the only scattering where mobility decreases with an increase in temperature owing to an increase in the number of phonons as the temperature goes up.

2.7 RELEVANCE OF HIGH-*k* GATE DIELECTRICS WITH TFET

To be adaptable with the latest nanotechnology, we need to lower the fabrication cost and simultaneously raise the device speed; therefore, conventional MOSFETs require replacement. TFET (Tunnel FET) offers various advantages in terms of steeper subthreshold slope and low off-current, and are proposed by various research groups [26, 27] as a substitute for application(s) wherein low standby power dissipation is of prime importance. It should be noted here that unlike MOSFET, constant field scaling does not fit to TFET [27], as conduction operation in TFET is totally unlike that of MOSFET. Kathy Boucart et al. have compared the effect of length scaling of TFET using various gate dielectrics that is SiO_2 and other high-*k* dielectrics like HfO_2/ZrO_2. They observed that TFETs have better capacitive coupling, and moreover, during scaling, it leads to better results. The bottom line is that FETs utilizing high-*k* dielectrics may be scaled down to shorter gate length before the OFF state becomes tainted by threshold voltage, off-current and sub-threshold swing. Further, pertaining to maximum electric field, due care is required in the ON state as well to evade reliability problems.

2.8 TFET

TFET, also referred to as surface tunnel transistor (STT) or Tunneling FET [28], is a p-i-n diode with reverse biasing that uses band gap as an energy barrier and works on the precept of band-to-band tunneling, which results in higher performance at low voltages in comparison to conventional MOSFETs.

The fundamental difference between the conventional MOSFET and TFET is the carrier injection mechanism. In MOSFETs, thermionic emission is used as a source of carrier injection mechanism, whereas TFET exploits band-to-band tunneling as a source of carrier injection mechanism.

Since we need to shoot down V_{DD} to keep switching energy low, which is not possible in conventional CMOS, when supply voltage is scaled down to 0.5 V, performance suffers significantly. Therefore, to overcome it, TFET offers:

    a. Sharper turn on devices in comparison to MOSFETs.
    b. Better performance at ultra-low power applications.

2.8.1 TFET Device Structure and Operating Principle

The basic TFET structure is similar to a MOSFET except that the source and drain terminals of a TFET are doped of opposite types (Figure 2.4). A common TFET device consists of a p-i-n junction (p-type, intrinsic, n-type) wherein the electrostatic potential of the intrinsic region is controlled by the gate terminal.

In order to understand the operating principle of TFET, it is necessary to compare it with a MOSFET. Figure 2.4 depicts the device structure for MOSFET and TFET. Double-gate structures are used here for illustration purposes only.

TFET mainly works on the principle of band-to-band tunneling. In TFETs, for band-to-band tunneling to ensue, an electron in the valence band of semiconductor must tunnel across the band gap to the conduction band without the aid of traps. The vital conditions for tunneling to occur are:

    a. The barrier must be fairly narrow over a large area for effectual tunneling.
    b. The density of states should be sufficient on both sides (i.e., source side as well as drain side) to offer energetic sites for the carriers.

2.8.2 Applications of TFET

The TFET fits in to the family of so-called steep slope devices that are utilized extensively in ultra-low-power electronic applications like digital switch, etc. Owing to their negligible off-currents, these devices are an apt choice for low standby power logic applications working at moderate frequencies. Besides these, the further applications include ultra-low-power specific analog ICs with improved temperature strength and

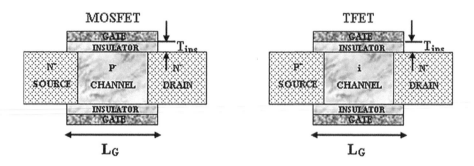

FIGURE 2.4   Basic structure of MOSFET and TFET.

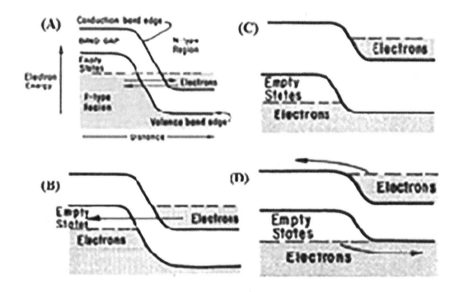

FIGURE 2.5   Energy band diagram for MOSFET and TFET.

low-power memories like DRAM and SRAM. Further, its application with high-*k* materials makes it a more promising candidate for future technology nodes.

2.9   APPLICATIONS OF HIGH-*k* MATERIALS

Besides gate oxides, high-*k* materials have an array of applications like:

     a. MEMS
     b. Coating materials
     c. Various nanotechnology and nano-electronic applications
     d. Electroluminescence
     e. As a passivation layer for Cu interconnects
     f. Storage capacitor dielectrics

2.10 SUMMARY

This chapter reviews the replacement of SiO_2 as the gate oxide in CMOS-based FET devices in length by analyzing various materials chemistry, bonding, electrical behavior, etc. The new oxide must adhere to the referred 10 conditions to be acceptable as gate dielectrics, namely large band gap, large *k* value, thermodynamic stability, long lifetime, low high-*k* interface surface density, stable morphology after heat treatment, high channel carrier mobility, low oxygen diffusion coefficients, large energy band gap, and good kinetic stability, besides few electrically active defects. The defects like oxygen vacancies, electron trapping, interstitials, remote charge scattering, etc. for the oxide must be optimized in order to attain high performance in devices. Basic physics behind advance FET device like TFET has also been reviewed. Consequently, it is observed that TFET devices along with high-*k* materials can be a most promising candidate for future low-power technology nodes.

ACKNOWLEDGMENTS

The authors would like to thank AICTE and NPIU, India for supporting this work with a grant under the project–collaborative research scheme, Project ID: 1-5741460991.

REFERENCES

[1] International Technology Roadmap for Semiconductors, "International Technology Roadmap for Semiconductors: Front End Processes," p. 23, 2003.

[2] E. J. Preisler, S. Guha, M. Copel, N. A. Bojarczuk, M. C. Reuter, and E. Gusev, "Interfacial oxide formation from intrinsic oxygen in W-HfO$_2$ gated silicon field-effect transistors,"*Appl. Phys. Lett.*, vol. 85, no. 25, pp. 6230–6232, 2004, doi:10.1063/1.1834995.

[3] K. Y. Tse and J. Robertson, "Control of schottky barrier heights on high-K gate dielectrics for future complementary metal-oxide semiconductor devices," *Phys. Rev. Lett.*, vol. 99, no. 8, pp. 2–5, 2007, doi:10.1103/PhysRevLett.99.086805.

[4] A. Callegari *et al.*, "IBM Research Report Electron Mobility Temperature Dependence of W/HfO$_2$ Gate Stacks: The Role of the Interfacial Layer," vol. 23700, 2005.

[5] Y. Y. Lebedinskii, A. Zenkevich, E. P. Gusev, and M. Gribelyuk, "In situ investigation of growth and thermal stability of ultrathin Si layers on the HfO$_2$/Si (100) high-κ dielectric system," *Appl. Phys. Lett.*, vol. 86, no. 19, pp. 1–3, 2005, doi:10.1063/1.1923158.

[6] M. R. Visokay, J. J. Chambers, A. L. P. Rotondaro, A. Shanware, and L. Colombo, "Application of HfSiON as a gate dielectric material," *Appl. Phys. Lett.*, vol. 80, no. 17, pp. 3183–3185, 2002, doi:10.1063/1.1476397.

[7] M. A. Quevedo-Lopez, J. J. Chambers, M. R. Visokay, A. Shanware, and L. Colombo, "Thermal stability of hafnium-silicate and plasma-nitrided hafnium silicate films studied by Fourier transform infrared spectroscopy," *Appl. Phys. Lett.*, vol. 87, no. 1, pp. 85–88, 2005, doi:10.1063/1.1977184.

[8] G. D. Wilk, R. M. Wallace, and J. M. Anthony, "High-κ gate dielectrics: Current status and materials properties considerations," *J. Appl. Phys.*, vol. 89, no. 10, pp. 5243–5275, 2001, doi:10.1063/1.1361065.

[9] A. I. Kingon, J. P. Maria, and S. K. Streiffer, "Alternative dielectrics to silicon dioxide for memory and logic devices," *Nature*, vol. 406, no. 6799, pp. 1032–1038, 2000, doi:10.1038/35023243.

[10] G. Pant *et al.*, "Effect of thickness on the crystallization of ultrathin HfSiON gate dielectrics," *Appl. Phys. Lett.*, vol. 88, no. 3, pp. 1–3, 2006, doi:10.1063/1.2165182.

[11] J. Robertson, "High dielectric constant gate oxides for metal oxide Si transistors," *Reports Prog. Phys.*, vol. 69, no. 2, pp. 327–396, 2006, doi:10.1088/0034-4885/69/2/R02.

[12] M. Houssa *et al.*, "Electrical properties of high-κ gate dielectrics: Challenges, current issues, and possible solutions," *Mater. Sci. Eng. R Reports*, vol. 51, no. 4–6, pp. 37–85, 2006, doi:10.1016/j.mser.2006.04.001.

[13] X. Zhang, A. Demkov, H. Li, X. Hu, Y. Wei, and J. Kulik, "Atomic and electronic structure of the Si/SrTiO₃ interface," *Phys. Rev. B – Condens. Matter Mater. Phys.*, vol. 68, no. 12, pp. 1–6, 2003, doi:10.1103/PhysRevB.68.125323.

[14] H. R. Huff and D. C. Gilmer, *High Dielectric Constant Materials: VLSI MOSFET Applications*. Berlin Heidelberg: Springer, 2005.

[15] E. P. Gusev *et al.*, "Ultrathin high-K metal oxides on silicon: Processing, characterization and integration issues," *Microelectron. Eng.*, vol. 59, no. 1–4, pp. 341–349, 2001, doi:10.1016/S0167-9317(01)00667-0.

[16] R. Ludeke, V. Narayanan, E. P. Gusev, E. Cartier, and S. J. Chey, "Potential imaging of Si/HfO₂/polycrystalline silicon gate stacks: Evidence for an oxide dipole," *Appl. Phys. Lett.*, vol. 86, no. 12, pp. 1–3, 2005, doi:10.1063/1.1890483.

[17] G. Bersuker *et al.*, "The effect of interfacial layer properties on the performance of Hf-based gate stack devices," *J. Appl. Phys.*, vol. 100, no. 9, 2006, doi:10.1063/1.2362905.

[18] B. H. Lee, S. C. Song, R. Choi, and P. Kirsch, "Metal Electrode/High-k Dielectric Gate-Stack Technology for Power Management," vol. 55, no. 1, pp. 8–20, 2008.

[19] J. C. Ranuárez, M. J. Deen, and C. H. Chen, "A review of gate tunneling current in MOS devices," *Microelectron. Reliab.*, vol. 46, no. 12, pp. 1939–1956, 2006, doi:10.1016/j.microrel.2005.12.006.

[20] J. Robertson and R. M. Wallace, "High-K materials and metal gates for CMOS applications," *Mater. Sci. Eng. R Reports*, vol. 88, pp. 1–41, 2015, doi:10.1016/j.mser.2014.11.001.

[21] M. H. Asghar, F. Placido, and S. Naseem, "Characterization of reactively evaporated TiO₂ thin films as high," *Eur. Phys. Journal Applied Phys.*, vol. 184, no. 3, pp. 177–184, 2006, doi:10.1051/epjap.

[22] J. F. Damlencourt *et al.*, "Study of HfO₂ films deposited on strained SI 1-xGex layers by atomic layer deposition," *J. Appl. Phys.*, vol. 96, no. 10, pp. 5478–5483, 2004, doi:10.1063/1.1805184.

[23] C. Maunoury *et al.*, "Chemical interface analysis of as grown HfO2 ultrathin films on SiO₂," *J. Appl. Phys.*, vol. 101, no. 3, pp. 2–8, 2007, doi:10.1063/1.2435061.

[24] G. Lucovsky, Y. Wu, H. Niimi, V. Misra, and J. C. Phillips, "Bonding constraints and defect formation at interfaces between crystalline silicon and advanced single layer and composite gate dielectrics," *Appl. Phys. Lett.*, vol. 74, no. 14, pp. 2005–2007, 1999, doi:10.1063/1.123728.

[25] S. J. Clark, J. Robertson, S. Lany, and A. Zunger, "Intrinsic defects in ZnO calculated by screened exchange and hybrid density functionals," *Phys. Rev. B - Condens. Matter Mater. Phys.*, vol. 81, no. 11, pp. 1–5, 2010, doi:10.1103/PhysRevB.81.115311.

[26] P. F. Wang *et al.*, "Complementary tunneling transistor for low power application," *Solid. State. Electron.*, vol. 48, no. 12, pp. 2281–2286, 2004, doi:10.1016/j.sse.2004.04.006.

[27] K. K. Bhuwalka, J. Schulze, and I. Eisele, "Scaling the vertical tunnel FET with tunnel bandgap modulation and gate workfunction engineering," *IEEE Trans. Electron Dev.*, vol. 52, no. 5, pp. 909–917, 2005, doi:10.1109/TED.2005.846318.

[28] A. C. Seabaugh and Q. Zhang, "Seabaugh_IEEE_Proceedings_review.pdf," vol. 98, no. 12, 2010.

3 Influence of High-*k* Material in Gate Engineering and in Multi-Gate Field Effect Transistor Devices

C. Usha and P. Vimala

Dayananda Sagar College of Engineering, Bengaluru, India

CONTENTS

3.1 INTRODUCTION: BACKGROUND AND DRIVING FORCES

In 1963, complementary metal oxide semiconductor technology (CMOS) was invented. CMOS technology is based on the organization of two metal oxide semiconductor field effect transistors (MOSFETs), one of which is a p-MOS and other an n-MOS. It is used for construction of electronic devices, integrated circuits, memories, microprocessors, and microcontrollers, among others. The CMOS technology has various advantages with respect to high package density, improved performance of the device when scaled down, low power consumption, etc. Gorden Moore made a prediction that the number of transistors on the chip doubles every 18 months [1]. Based on this prediction, the devices are scaled down from the millimeter to nanometer dimension. The reduction in the size of devices is analyzed by International Technology Roadmap for Semi-conductors (ITRS) for year 2020 [2] as shown in the Figure 3.1.

DOI: 10.1201/9781003121589-3

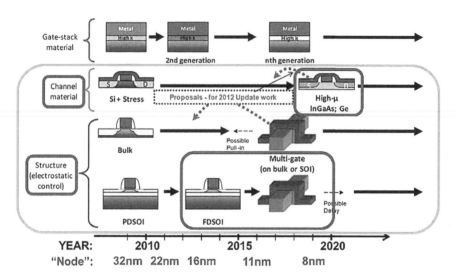

FIGURE 3.1   International technology road map for semiconductors for year 2020 [2].

MOSFET devices have been developed based on the scaling law. MOSFET is an insulated gate field effect transistor and voltage-controlled device. The basic MOSFET planar structure is shown in Figure 3.2 [3]. The structure consists of a single gate with p-type substrate uniformly doped and gate oxide, source, and drain doped with n-type material. The dimension and features of the planar device were reduced for the requirements of high packaging density, low power consumption, lower propagation delay, and higher-frequency operations. Based on the type of doping at source and drain, MOSFET is named as n-MOS for n-type of doping and p-MOS for p-type of doping.

Two types of MOSFETs are classified based on mode of operation: one is the depletion type and the other is the enhancement type. Figure 3.3 shows the device structure of enhancement and depletion mode of p-MOS and n-MOS transistors. Depletion-mode MOSFET is correspondent to the normally switch OFF condition, and the enhancement mode is correspondent to the normally switch ON condition.

In n-type MOSFET, the substrate is of p-type with moderately doped source and drain regions, which are diffused with the n-type impurity for the formation of the depletion region. The source and drain connections are made through the metal deposited. The electrons are the charge carriers for the conduction. The current across the channel is controlled in one of the two different ways: enhancement or depletion. Similarly, in p-type MOSFET, the substrate is of n-type with moderately doped source and drain regions, which are diffused with the p-type impurity. The holes are the charge carriers for the conduction in p-type MOSFET. As the device is scaled down, short channel effects (SCEs) arise that lead to reduced performance of the device. Short channel influences are more prominent when the length of the channel comes within the width of the junction. To reduce the SCEs and increase the performance of the device, dual-gate or double (DG) MOSFET and gate-all-around (GAA) or surrounding- gate MOSFETs are used.

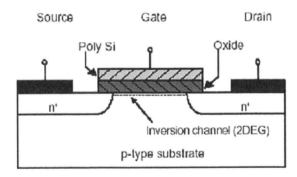

FIGURE 3.2   Single-gate MOSFET structure [3].

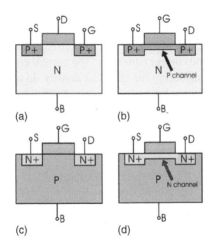

FIGURE 3.3   (a) Enhancement mode p-type MOSFET, (b) Depletion mode p-type MOSFET, (c) Enhancement mode n-type MOSFET, (d) Depletion mode n-type MOSFET.

3.2   DOUBLE-GATE METAL OXIDE SEMICONDUCTOR FET (DGMOSFET)

Double-gate MOSFET is one of the important devices of multi-gate MOSFET family. DGMOSFET has improved switching characteristics and high transconductance compared to single-gate MOSFETs, which enables scalability to sub-50 nm VLSI/ Ultra Large-Scale Integration applications.

Figure 3.4 shows the DGMOSFET structure consisting of two gates, with the top and bottom gates of same thickness. Based on the material used for gates, DGMOSFET is categorized into two types: symmetric DGMOSFET and asymmetric DGMOSFET. In symmetric DGMOSFET the material used for both gates is the same. In asymmetric DGMOSFET the materials with different work functions are used for the top gate and the bottom gate, respectively. DG MOSFET structure

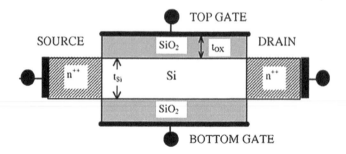

FIGURE 3.4 Schematic structural view of DGMOSFET [4].

realized on thin films increases the number of electrons and mobility due to volume accumulation or inversion, which further drives the high drain current and reduces the SCEs and OFF-state current.

The DGMOSFET can be modeled using the classical and quantum mechanical calculations. In classical calculations, the self-consistent solution is obtained from the Poisson's equation and current equations. The current equation is obtained by the drift-diffusion model considering the classical density. In quantum mechanical calculation, the self-consistent solution is gained by Poisson's, Schrodinger's, and current equations. The current equation can be obtained by the drift-diffusing model considering quantum mechanical density. The conduction band and electron density variation diagram related to the classical and quantum mechanical modeling are shown in Figure 3.5.

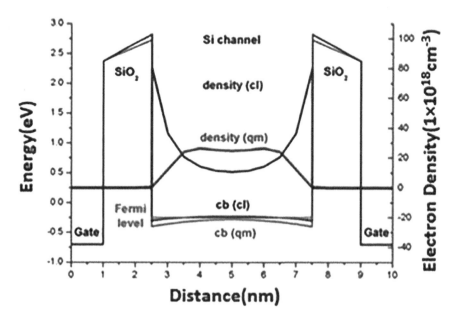

FIGURE 3.5 Conduction band and electron density variation along the DGMOSFET device [5].

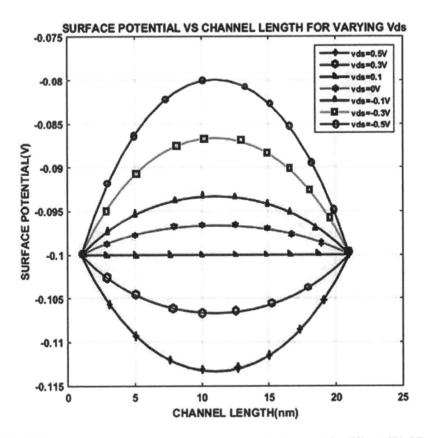

FIGURE 3.6 Surface potential variation across the length of channel for different Vds [6].

From Figure 3.5 it is observed that the Fermi level of both models, classical and quantum mechanical, is flat. The conduction band edge across the channel region in the quantum mechanical is lower compared to the classical. Electron density across the silicon and silicon dioxide interface is maximum in the classical and zero in the quantum mechanical analysis. The surface potential in the device is due to change in potential from doped to intrinsic. The surface potential distribution across the length of the channel is shown in Figure 3.6. The electrical field distribution of two different types occurs in the device shown in Figure 3.7. The lateral electrical field occurs from nonzero source to the drain region; this mechanism impacts the drain current. The vertical electrical field distribution is due to potential variation between the gate and the substrate.

3.3 CYLINDRICAL GATE-ALL-AROUND METAL OXIDE SEMICONDUCTOR FET (CGAA MOSFET)

In cylindrical GAA MOSFET the channel region is surrounded by the gate that has higher control over the channel, which drives higher current and further minimizes the size of the device and short channel effects. Due to dimension miniaturization

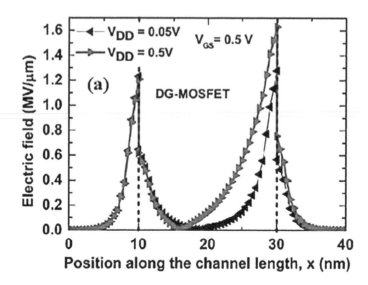

FIGURE 3.7   Electrical field distributions along the channel length [7].

and high current drive, CGAA MOSFET can accomplish the high packaging density compared to the DGMOSFET structure. Therefore, the CGAA MOSFET structure has exceptional electrostatic channel control, SCEs robustness, enhanced scaling options, greater equivalent number of gates, improved sub-threshold swing limit, reduced corner effects, carrier's non-confinement at Si/SiO₂ interface, and compact device length. Figure 3.8 shows the three-dimensional structural view of the CGAA MOSFET, in which the channel is fully surrounded by the gate.

The charge carriers in nanometer-scale CGAA MOSFET device are confined in two dimensions. Thus, potential distribution of the CGAA MOSFET is similar to the DG MOSFET, and the analysis of the device is carried out by considering the radius and z-axis of the cylindrical device. To understand the transport properties and electrostatic characteristics of the CGAA MOSFET, Poisson's equation solution is used. Two approximations can be carried out to understand the device characteristic when Poisson's equation is used. The simulation of cylindrical structure is obtained by finite element method. As the size of the device is reduced, the quantum effects in the simulation should be considered by the usage of self-consistent solution in Schrodinger equation. Figure 3.9 shows the potential variation of the CGAA MOSFET device; it is observed that the potential variation is high compared to the DGMOSFET device.

FIGURE 3.8   3D structural view of cylindrical gate-all-around MOSFET [8].

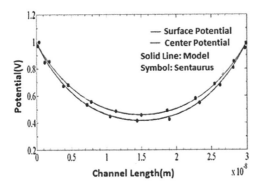

FIGURE 3.9   Potential variation across the length of channel.

3.4   GATE ENGINEERING OF MULTI-GATE MOSFET

The variation of the gate metal work function plays an important role in the gate engineering of devices. The work function is represented with difference of energy with lattice, which directly affects the drain current. The two and three different work function materials are used as a gate engineering technology for multi-gate devices. The gate engineering devices include dual-metal (DM) DG MOSFET, triple-metal (TM) DG MOSFET, dual-metal (DM) GAA MOSFET, and triple-metal (TM) GAA MOSFET.

3.4.1   DUAL-METAL DOUBLE-GATE MOSFET

In double-metal gate engineering, two metals with different work functions are used. The two metals used are combined together to provide instantaneous increase in drain current and transconductance and to suppress short channel effects when compared with the single-metal gate. Figure 3.10 depicts the schematic diagram of dual-metal (DM) DG MOSFET, in which the two different work functions ϕ_{m1} and ϕ_{m2} materials are used. The gate with work function ϕ_{m1} near to the source channel region represents the control gate. The gate near to drain channel region with work function ϕ_{m2} represents the screen gate of the device. In general, the control gate metal with high work function is chosen when compared to the screen gate metal. Due to this selection, the electron charge velocity and lateral electrical field at interface of two metals across the channel increase, which in turn increases the gate transport efficiency.

The DMDG MOSFET can be modeled using classical and quantum mechanical modeling. The comparison plot of surface potential profile for DG MOSFET and DMDG MOSFET is depicted in Figure 3.11. The plot shows the surface potential profile change across the two metal gates. This change is due to the use of two different work function metals in the gate. It is also observed that the potential of the DMDG MOSFET is improved compared to the DG MOSFET. Figure 3.12 shows the electric field comparison plot of DMDG MOSFET and DG MOSFET, and it is observed that the electrical field peak is reduced when compared with DG MOSFET due to increased carrier transport velocity. Figure 3.13 shows the drain current variation, and it is observed that drain current variation of DMDG MOSFET is high compared to DG MOSFET due to improved electric field.

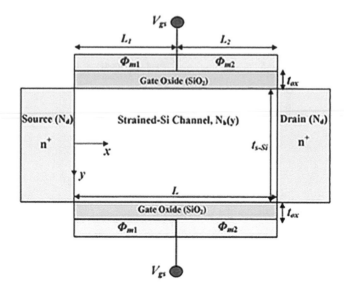

FIGURE 3.10   Schematic diagram of dual-metal double-gate (DMDG) MOSFET [9].

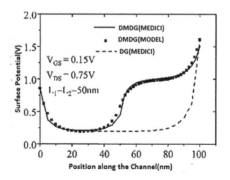

FIGURE 3.11   Surface potential comparison of single-material DGMOSFET and dual-material DGMOSFET [10].

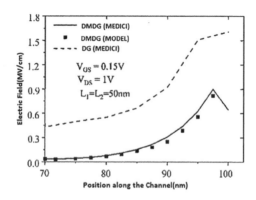

FIGURE 3.12   Electric field comparison plot of DMDG MOSFET and DG MOSFET [10].

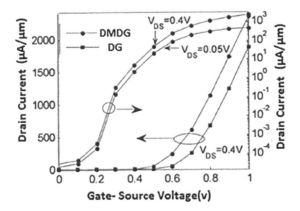

FIGURE 3.13 Comparison plot of drain current with gate to source voltage of SMDG and DMDG MOSFET [16].

3.4.2 TRIPLE-MATERIAL DOUBLE-GATE MOSFET

Similar to double material, in triple-material gate engineering, three metals with different work function are used at gate. The use of three metals further improves the electrostatic characteristics of the device and suppresses the short channel effects. Figure 3.14 shows the schematic diagram of triple-material (TM)

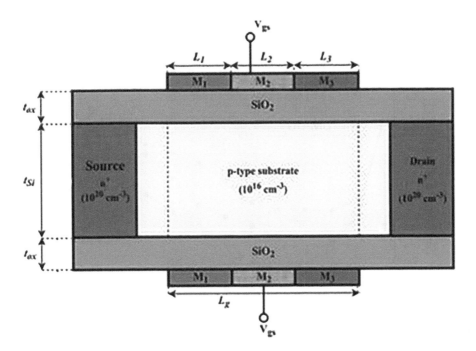

FIGURE 3.14 Schematic diagram of triple-metal DG MOSFET.

DGMOSFET. The three materials are used for gate engineering. Figure 3.15 presents the comparison plot of surface potential distribution of DG MOSFET, DMDG MOSFET, and TMDG MOSFET. It is observed that the potential distribution varies across the two interfaces of metals used across the gate and improved surface potential compared to DMDG MOSFET. The lengths of the metals can also be varied for further improvement in electrostatic characteristics. The ratio of metal lengths variations has showed higher improvement than the constant ratio of metal lengths.

Figure 3.16 shows the comparison plot of lateral electric field for DG MOSFET, DMDG MOSFET, and TMDG MOSFET. This plot depicts two electric field peaks in TMDG MOSFET across the length of the channel in DMDG MOSFET. From the plot it is observed that the lateral electrical field persuaded by the additional voltage at drain is fully absorbed at the screening second metal gate M2 and third metal gate M3, which reduces the electrical field level at the drain side. Due to this, hot carrier effects (HCEs) decrease and average lifetime of the device increases. The additional peak in graph indicates the increase in the carrier mobility and increase in efficiency of carrier transport at the interface of gate metal. This warrants improved transconductance, drain current, and drain induced barrier lowering (DIBL) values in TMDG MOSFET device [11,12]. The improvement in the transconductance in the TMDG MOSFET is shown in Figure 3.17. The drain current of the TMDG MOSFET is shown in Figure 3.18. The drain current plotted indicates the higher current compared to the DMDG MOSFET due to higher electron mobility with respect to three materials used.

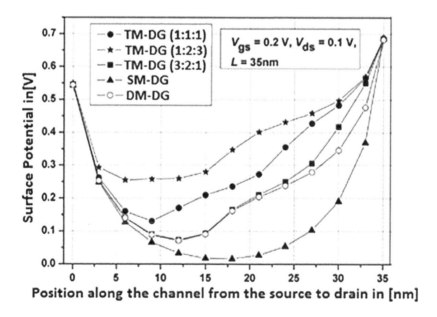

FIGURE 3.15   Comparison plot of surface potential distribution across the channel length of SMDG, DMDG, and TMDG MOSFET.

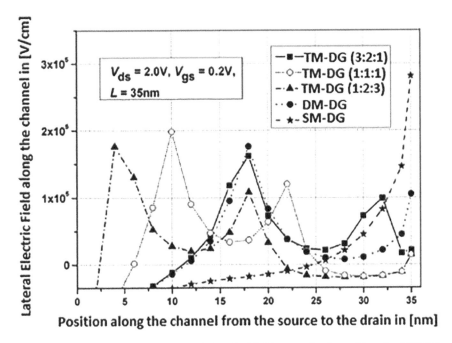

FIGURE 3.16 Comparison plot of lateral electrical field across the channel length of SMDG, DMDG, and TMDG MOSFET.

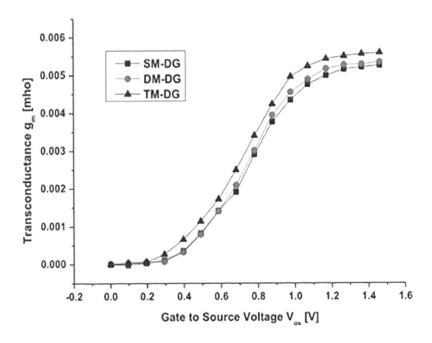

FIGURE 3.17 Comparison plot of the transconductance with gate to source voltage of SMDG, DMDG, and TMDG MOSFET [12].

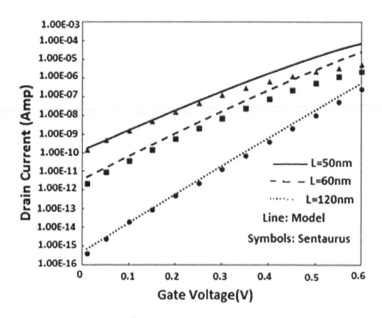

FIGURE 3.18   Drain current with gate to source voltage of TMDG MOSFET.

3.4.3   Double-Metal Surrounding-Gate (DMSG) MOSFET

Surrounding-gate (SG) MOSFETs have proven to be popular for nanoscale structures by providing an enhanced scalability selection along with reduced SCEs, higher
transconductance, and near ideal sub-threshold slope. Combining the structural
advantages of nanowire MOSFET along with the gate engineering techniques advantages, a new structure is introduced called dual-metal surrounding-gate (DMSG)
MOSFET [13]. The DMSG MOSFET can be modeled on Poisson's equation using
two approximation methods: parabolic approximation and superposition approximation. As the size of the device is further reduced to nanoscale, the quantum effects are
considered and modeled using Schrodinger's equation. Figure 3.19 shows the

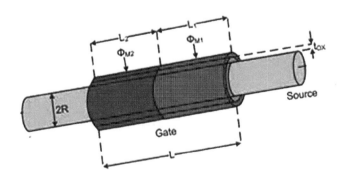

FIGURE 3.19   Three-dimensional view of dual-metal surrounding-gate MOSFET [14].

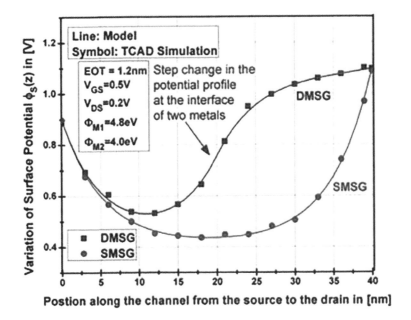

FIGURE 3.20 Comparison plot of surface potential profile across the length of the channel for SMSG and DMSG MOSFET [14].

three-dimensional view of the DMSG MOSFET; two metals with different work function are used, with two lengths of metal L_1 and L_2.

Figure 3.20 depicts the comparison plot of surface potential variation across the channel; as the DMSG MOSFET gate is a combination of two different work function metals, a step potential variation occurs across the two different work function junction. This potential variation indicates the increase in carrier velocity, increase in carrier transport efficiency, and improvement in drain current. Figure 3.21 depicts the comparison of lateral electrical field along the length of channel. The step potential change across the two metals junction causes the lateral electrical field peak change. This occurrence of the peak change represents the increase in carrier transport efficiency.

3.4.4 TRIPLE-METAL SURROUNDING-GATE (TMSG) MOSFET

Gate engineering increases the electrostatic characteristics of the device. Thus, three metals with different work function are used for the gate terminal. Figure 3.22 shows a schematic view of TMSG MOSFET, in which each metal length is considered by different lengths of L_1, L_2, and L_3. For better performance, the higher work function of the metal at the source side is considered compared to the metal used at the drain side. Similar to the DMSG MOSFET, TMSG MOSFET is modeled on the Poisson's equation using the two approximation methods: parabolic approximation and superposition approximation. As the size of the device is further reduced to nanoscale, the quantum effects are considered and modeled using Schrodinger's equation.

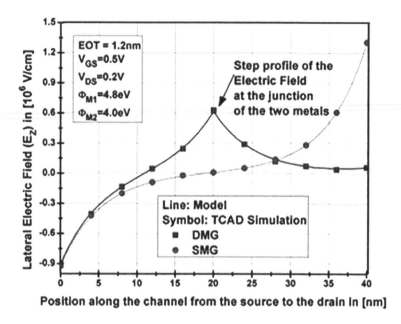

FIGURE 3.21 Comparison plot of lateral electrical field variation along the length of channel for SMSG and DMSG MOSFETs.

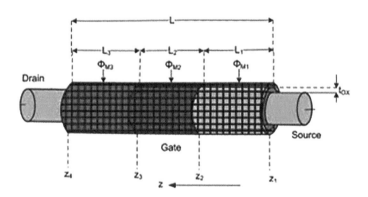

FIGURE 3.22 Three-dimensional view of TMSG MOSFET.

Comparison of surface potential profile for the SMSG, DMSG, and TMSG MOSFETs is shown in the Figure 3.23. The step potential change occurs across the two junctions of the different work function metals. This change improves carrier velocity, carrier transport efficiency, improves drain current. The electrical field profile comparison plot is shown in Figure 3.24. Due to step potential change across the

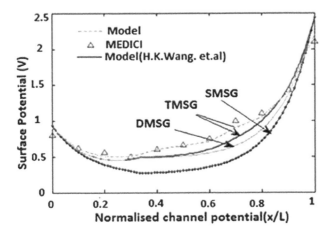

FIGURE 3.23 Comparison plot of surface potential profile across the normalized length of channel for SMSG, DMSG, and TMSG MOSFETs.

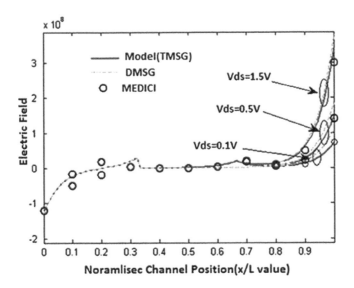

FIGURE 3.24 Comparison of electrical field profile across the normalized length of channel for SMSG, DMSG, and TMSG MOSFETs.

two junction potentials, the electric field also varies. This variation of the electric field profile increases carrier transport efficiency. Because of higher electrical field profile, the drain current increases as shown in Figure 3.25.

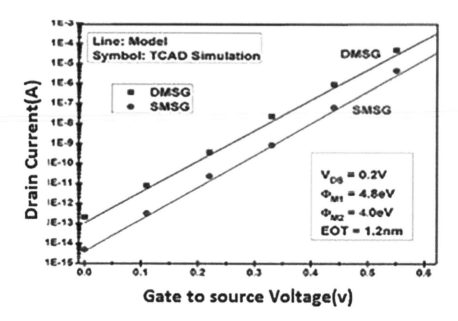

FIGURE 3.25   Comparison plot of drain current with gate to source voltage of DMSG and TMSG MOSFET.

3.5   IMPACT OF HIGH-*k* DIELECTRICS IN GATE ENGINEERING AND MULTI-GATE MOSFET STRUCTURES

As the size of the multi-gate MOSFET devices is reduced, the silicon dioxide layer is reduced to increase gate capacitance and to drive higher drain current. Due to this decrease in silicon dioxide layer, the tunneling increases drastically, which leads to leakage currents, high power consumption, and reduced device reliability. The use of high-*k* dielectric increases gate capacitance and restricts leakage current. The high-*k* dielectric materials with high permittivity enable the same thickness of electrical oxide layer thicker than the silicon oxide layer. This decreases the leakage across the gate dielectric. In addition, metal gates in the high-*k* dielectric will dynamically screen electrons. This raises the movement of charge carriers along the channel. Because of this advantage, the high-*k* can be stacked in gate engineering and multi-gate MOSFET devices.

The materials with high permittivity deposited using atomic layer deposition for devices are Si_3N_1, Al_2O_3, $LaAlO_3$, HfO_2/ZrO_2, and La_2O_3, which are used for the multi-gate MOSFET devices due to high performance of electrostatic characteristic compared to other materials. The high-*k* stacked in gate engineering and multi-gate MOSFETs are high-*k* stacked double-gate MOSFET, high-*k* stacked DMDG MOSFET, high-*k* stacked TMDG MOSFET, high-*k* stacked gate-all-around MOSFET, high-*k* stacked DMSG MOSFET, and high-*k* stacked TMSG MOSFET. The structures of high-*k* stacked in gate engineering and multi-gate MOSFET are shown in Figure 3.26. The high-*k* stacked surrounding-gate MOSFET modeling is based on the cross-sectional view of the device, such that the schematic view of the surrounding gate

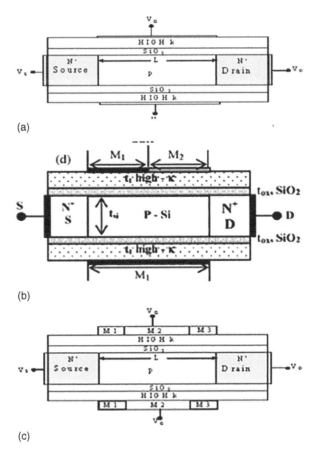

FIGURE 3.26   Schematic diagram of (a) high-*k* stacked DG MOSFET, (b) high-*k* stacked DMDG MOSFET, (c) high-*k* stacked TMDG MOSFET.

device is same. Figure 3.27 shows the surface potential variation comparison plot of the high-*k* stacked DG MOSFET and DMDG MOSFET [15]. It is observed that due to high *k*, the surface potential profile is high compared to without high-*k* stacked devices. Electric field profile of high-*k* stacked DG and DMDG MOSFET is shown in Figure 3.28. This is due to high carrier transport in the device. This in turn increases the drain current as shown in Figure 3.29. Due to the impact of high *k*, the transconductance improvement is shown in Figure 3.30.

Figure 3.31 presents the comparison plot of TMDG MOSFET with addition of high *k* and without addition of high *k*; it is posited that the gate engineering and high-*k* inclusion have further improved the performance of the device with respect to electrostatic potential. Figure 3.32 depicts the comparison plot of electric field profile; it is suggested that as surface potential is high, the electrical field is also high for the high-*k* stacked TMDG MOSFET. Increased electric field profile further improves the drain current and reduces the leakage current. Figure 3.33 shows the drain current variation for with and without high-*k* stacked TMDG MOSFET. The drain current for high-*k* stacked devices is high compared to those without due to higher tunneling of electrons.

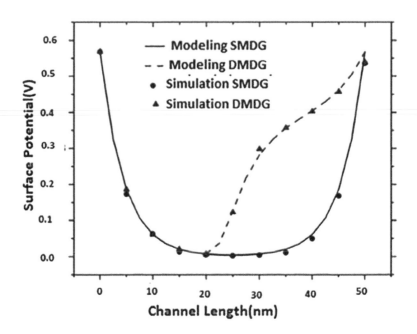

FIGURE 3.27 Comparison plot of surface potential profile for high-*k* stacked SMDG and high-*k* stacked DMDG MOSFET [14].

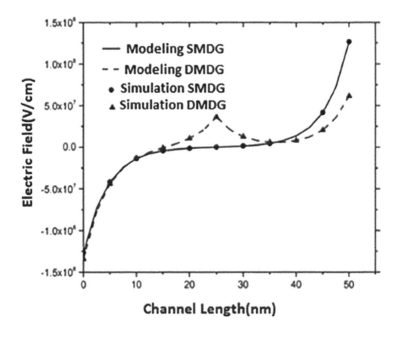

FIGURE 3.28 Comparison plot of electric field profile for high-*k* stacked SMDG and high-*k* stacked DMDG MOSFET [14].

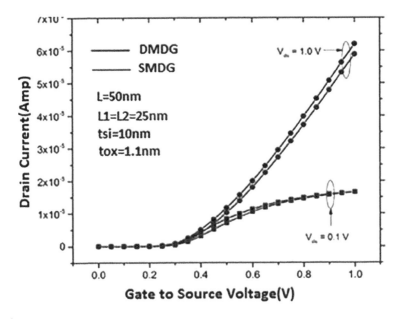

FIGURE 3.29 Comparison plot of drain current for high-*k* stacked SMDG and high-*k* stacked DMDG MOSFET [14].

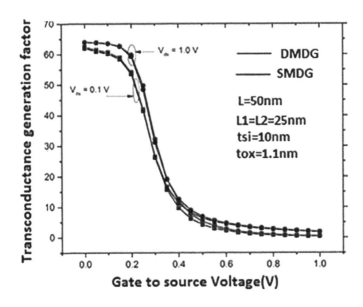

FIGURE 3.30 Comparison plot of transconductance for high-*k* stacked SMDG and high-*k* stacked DMDG MOSFET [14].

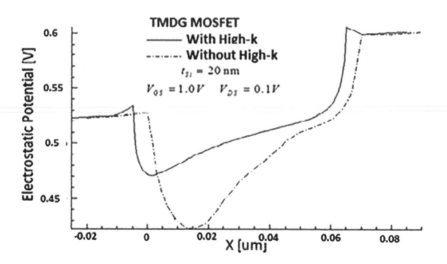

FIGURE 3.31   Electrostatic potential plot of TMDG MOSFET without and with high-*k* stacked [17].

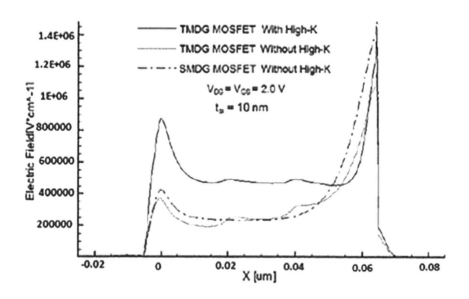

FIGURE 3.32   Electrical field profile of TMDG MOSFET without and with high-*k* stacked [17].

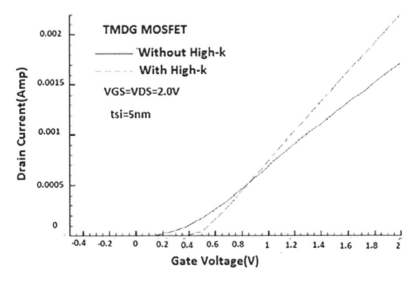

FIGURE 3.33 Drain current variation for TMDG MOSFET without and with high-*k* stacked [17].

REFERENCES

[1] G. Moore. 1955. Cramming more components onto integrated circuits. *Electronics* 38: 529–551.

[2] International Technology Roadmap for Semiconductors. 2011 Edition.

[3] F. Schwierz, and J. J. Liou. 2003. *Modern Microwave Transistors: Design, Modeling and Performance*. Hoboken: Wiley.

[4] A. Rahman, and M. S. Lundstrom. 2015. Erratum: A compact scattering model for the nanoscale double-gate MOSFET. *IEEE Transactions on Electron Devices* 62: 2367.

[5] S. Birnera, S. Hackenbuchnera, M. Sabathila et al. 2006. Modeling of Semiconductor Nanostructures with nextnano. *Acta Physica Polonica A* 110: 111–124.

[6] P. Vimala and N.R.N. Kumar. 2018. Performance analysis of the channel double gate MOSFETs with high K/Ge material based on inversion charge model. *Journal of Nanotechnology & Advanced Materials* 6(2): 21–32.

[7] J. Arin, J. Akhi, S. Azam, and A. K. Ajad. 2019. GaN-based double gate-junctionless (DG-JL) MOSFET for low power switching applications. *Materials Science 2019 International Conference on Electrical, Computer and Communication Engineering (ECCE)*. 1–4.

[8] R. Hosseini, M. Fathipour, and R. Faez. 2012. Quantum simulation study of gate-all-around (GAA) silicon nanowire transistor and double gate metal oxide semiconductor field effect transistor (DG MOSFET). *International Journal of the Physical Sciences* 7: 5054–5061.

[9] E. Goel, K. Singh, S. Kumar, and S. Jit. 2017. 2-D analytical modeling of subthreshold current and subthreshold swing for ion-implanted strained-Si double-material double-gate (DMDG) MOSFETs. *Indian Journal of Physics* 91: 1069–1076.

[10] G. Venkateshwar Reddy, and M. Jagadesh Kumar. 2005. A new dual-material double-gate (DMDG) nanoscale SOI MOSFET—two-dimensional analytical modeling and simulation. *IEEE Transactions on Nanotechnology* 4: 260–268.

[11] Md. S. Rahman, and N. Ahmed. 2018. Impact of gate underlap design on analog and RF performance for 20nm tri-material double gate (TMDG) MOSFET. *Conference*.

[12] R. Ramesh. 2017. Influence of gate and channel engineering on multigate MOSFETs-A review. *Microelectronics Journal* 66: 136–154.

[13] M. J. Kumar, A. A. Orouji, and H. Dhakad. 2006. New dual-material SG nanoscale MOSFET: Analytical threshold voltage model. *IEEE Transactions on Electron Devices* 53: 920–922.

[14] P. Dhanaselvam, and N. Balamurugan. 2013. Analytical approach of a nanoscale triple-material surrounding gate (TMSG) MOSFETs for reduced short-channel effects. *Microelectronics Journal* 44: 400–404.

[15] V. Narendar, and K. A. Girdhardas. 2018. Surface potential modeling of graded-channel gate-stack (GCGS) high-K dielectric dual-material double-gate (DMDG) MOSFET and analog/RF performance study. *Silicon Journal* 10: 2865–2875.

[16] Z. Arefinia, and A. A. Orouji. 2018. Quantum simulation study of dual-material double gate (DMDG) MOSFET: NEGF approach. *IEEE Silicon Nanoelectronics Workshop* 1–4.

[17] S. K. Gupta, A. Baidya, and S. Baishya. 2012. Simulation and analysis of gate engineered triple metal double gate (TM-DG) MOSFET for diminished short channel effects. *International Journal of Advanced Science and Technology* 38: 15–24.

4 Trap Charges in High-*k* and Stacked Dielectric

Annada Shankar Lenka and
Prasanna Kumar Sahu
National Institute of Technology, Rourkela, India

CONTENTS

4.1 INTRODUCTION

The growing demand for electronic products in almost every part of life always encourages the semiconductor industries and researchers to discover new technologies, designs, or optimization of older ones for better operation of transistors. Simultaneously, the overall small product size, lower power consumption, and most importantly the reduction of short channel effects (SCEs) force us to redirect our perspective toward technological development. In other words, we can say that the control of electrical properties in every measure of the operation of a transistor or metal oxide semiconductor (MOS) system is always the main objective of our research. In 1984, a proposed design for a double-gate MOSFET by Sekigawa and Hayasi [1] brought a new era in the field of semiconductor, which enhanced the transistor performance by introducing more gate control over the channel or majority charge carrier. This concept enlightens the path of transistor industries to introduce multiple-gate devices (MU-G), which nowadays is the most widely utilized concept for advanced non-classical CMOS devices (AnCD). The well-known devices such double gate, triple gate, enclosed gate, gate all around, and FINFET are all part of this group. Many researchers believe that the above-mentioned technology helps us achieve the golden rule of transistor industry, posited by Gordon Moore in a famous 1965 article: that the number of transistors per chip would quadruple in every three years [2]. But if we consider designing a multi-gate structure, we need to introduce a higher number of dielectric layers with respect to each number of gates so the trap charges creation or presence in these layers is obviously higher than the single-gate transistor, thus influencing the channel behavior like inversion layer built-up, threshold voltage roll-off, gate leakage current, etc. Again, in memory-based design, these dielectric materials some time have more relaxation current due to the defects created during the fabrication or operation. Fabrication of these layers is more complicated than the conventional SiO_2, and hence choosing appropriate material for specific application can be vital.

4.2   NEED FOR HIGH-*k* DIELECTRIC

The operational principle of MOS capacitance is completely dependent on the gate bias, and for many application purposes we have to bias gate throughout the operation of the product. Hence this also causes the increase in temperature of the device, which results in density of interface trapped and oxide trapped charges [3,4], and this phenomenon affects the carrier mobility and threshold voltage of the transistor. The common term we use for this is BIOS temperature instability (BTI) [5,6]. However, this limitation arises only for modern node technology where nano-range oxide sheets are used as gate dielectric and are not capable of sustaining even a small voltage shift and finally result in circuit failure [7].

Another major limitation that comes into the picture due to aggressive scaling of semiconductor devices is off-state gate leakage current, since the physically ultra-thin dielectric (SiO_2) layer is much more prone to tunneling [8,9]. But the material with high dielectric constant can be used instead of conventional SiO_2, which helps the manufacturing foundry maintain the physical thickness of the gate oxide layer as desirable [8–12] according to the application. Normally the materials having dielectric constants (ε_{high-k}) ranges from 10 to 40 are used as gate oxide materials in fabrication industries [12]. The common high-*k* dielectric materials used for the above purpose are aluminum oxide, hafnium dioxide, zirconium dioxide, titanium dioxide, and tantalum pentoxide, among others [10–18]. However, along with the dielectric constant we also have to consider another major affecting factor, namely that the material has lower barrier height due to a small band gap, leading to the leakage current increasing exponentially. Thus, in the process of design and fabrication, the manufacturer must consider a trade-off between dielectric constant and barrier height [19, 20].

We also must consider another promising factor for the fabrication point of view, specifically that in most cases, high-*k* dielectric materials do not establish a direct interface with the silicon substrate, for which we normally use an interfacial layer between the Si substrate and high-*k* oxide layer. This also provides some other benefits to the designers, such as increased capacitance and thickness, which in turn reduces tunneling probability and the chances of excess silicon oxidation during the fabrication process. Nitrogen atoms are usually injected to these layers through thermal annealing, which also causes excess trapping of charge carrier at the interface. Due to the high density of intrinsic defects present in the high-*k* material, the external field induces trap generation, and retention is more common than with the conventional SiO_2. The trap disturbs the channel's majority and minority carrier and thus affects the drive current and V_{th} instability. Trap charge exploration in transistor, designed for the area of application where the external radiation impact on the overall performance of the device, would be more important than the general-purpose application; in that case we need to focus on these trap charges more for the reliability and lifetime of the whole circuit. In this chapter we will discuss types of trap charges and their effect on both SiO_2 and high-*k* material (some of the important available research work in this field is related to how the negative bias temperature instability (NBTI) affects the charges present in the oxide region and also considers some of the extraction techniques specially designed for target dielectric).

4.3   BACKGROUND AND HISTORY OF RESEARCH

The topic we need to focus on first is the different type of trapped charges at interface, border, and at the bulk with the high-*k* material application as a dielectric layer. We also need to discuss the effect of these charges on the behavior of the operational device.

Ion radiation during the fabrication procedure or at the application area mainly enhances the chances of charge accumulation at the semiconductor and oxide interface. Electrons and holes (e^-/h^+) generated due to the external radiation first undergo recombination, and the rest number of charge, also known as charge yield, is always affected by the applied electric field. As is well known, due to high mobility of electron with respect to holes, electrons are more affected by the applied gate bias at conventional operating condition. However, some of the electrons as well as holes undergo recombination process with injected charge carriers by the channel region. Let's consider application of a constant and positive gate bias. For this condition, from the total free charge carriers, electrons will be attracted to the metal and dielectric interface and form a negative charge sheet, and holes will be repelled back toward the Si–oxide interface and form a positive charge sheet. Again, due to the presence of positive charges at the interface of Si and SiO_2, it helps to accumulate hydrogen ions, which creates a silicon hydrogen bond (Si-H). These phenomena lead to threshold voltage roll-off, decrease in carrier mobility, and increase in leakage current [21] (Figure 4.1).

Some high-*k* dielectrics can also trap electrons or negatively charged ions, but generally the net charge of the high-*k*/SiO_2 dielectric store is positive [22, 23]. We commonly consider that interface charges are neutral when the Fermi level is present at the mid-gap for mathematical calculation as well as design perspective [24, 25]. Hence, we can say that the entire mid-gap voltage fluctuation (ΔVmg) is due to trapping of charges at the bulk oxide and the C-V curve, but the stretch-out or slope at the curve completely depends on the edge trap (ΔV_{it}).

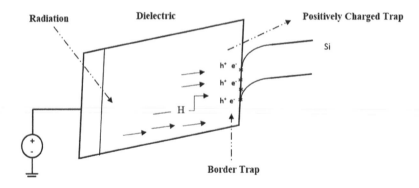

FIGURE 4.1   Energy band diagram of p-sub MOS structure showing different trap charges.

MOS transistor has a majority charge carrier as electrons and holes experience a positive and negative shifts in threshold voltage (V_{th}), respectively, due to interface confined charges [26]. But the oxide trap with net positive value forces V_{th} to shift toward more and more negative for both n- and p-channel MOS. So, if we consider only the n-substrate device, then we can see the total effect of both of the above charges on threshold voltage or we can say that it makes it more difficult to turn on the device. For mathematical calculation, the sub-threshold as well as C-V measurement gives the equivalent result to extract these charges with an assumption that across the band gap the interface charge density is not very significant.

By the help of mid-gap charge separation method we can determine the change in density of interface (ΔN_{it}) and oxide trap charges (ΔN_{ot}) [9, 26].

$$\Delta N_{it} = C_{ox}\left\{\left(\Delta V_{fb} - \Delta V_{mg} / qA\right)\right\} \tag{4.1}$$

$$\Delta N_{ot} = -C_{ox}\left\{\Delta V_{mg} / qA\right\} \tag{4.2}$$

where ΔV_{mg} is the mid-gap voltage shift and A is the area of the operating capacitor, and all other parameters carry the same known meaning and value. If we denote ΔV_{ot} and ΔV_{it} for voltage roll-off due to the oxide trap and interface trapped charges, respectively, then the next equations estimate the values of these.

$$\Delta V_{it} = \Delta V_{th} - \Delta V_{mg} \tag{4.3}$$

$$\Delta V_{ot} = \Delta V_{mg} \tag{4.4}$$

Now we will introduce another kind of trap charge into our discussion for a more detail analysis. The charged area presents in the dielectric and is able to communicate or exchange total net charge with the silicon body during the operation and is called a border trap. It was first proposed by Fleetwood in 1991 [27]. These types of charge sometimes relates to the defects in a crystal caused by a hydrogen ion at the edge of dielectric [28–31]. Due to some charge exchange property of these charges with interface trapped, the exact contribution and its effect on electrical behavior of the device is much more difficult (Figure 4.2). By the detailed observation of difference

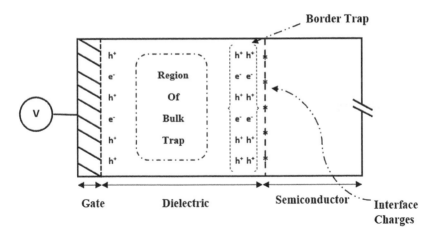

FIGURE 4.2 Position of different kind of trap charges in a MOS capacitor [28].

in capacitance from C-V curve hysteresis we can calculate the total density of border charges (ΔN_{bt}) [32].

$$\Delta N_{bt} = \left[\int \{C_{rev} - C_{for}\} dV\right]\left[\frac{1}{qA}\right] \tag{4.5}$$

$$\Delta V_{bt} = \Delta N_{bt}\left\{\frac{q}{C_{ox}}\right\} \tag{4.6}$$

4.4 REVIEW OF NEGATIVE BIOS TEMPERATURE INSTABILITY (NBTI)

As we discussed earlier, the NBTI condition arises when the gate is at negative bias for a prolonged time period and/or at higher operating temperature. Due to this, oxide as well as interface traps are created in the region. With older node technologies where we used high-thickness dielectric below the gate metal, NBTI was generally discussed with the H_2O or O_2 vacancies at the Si/dielectric interface [33–35].

The most common model used to describe the NBTI issue is diffusion-reaction (DR). This is because it uses a two-step equation like generation of defect charges and the area covered by the generated traps; the second is the effect of diffusion on the bulk silicon [36–37].

$$\text{Hole} + \text{Inactive defect charges}(\text{Positive}) \longleftrightarrow \text{Charged oxide sheet} \tag{4.7}$$
$$+ \text{ Interface traps}$$
$$+ \text{ Mobile ion at interface}$$

$$\text{Interface mobile ion} \longleftrightarrow \text{Bulk mobile ion}(\text{Diffusion Process}) \tag{4.8}$$

Usually the energy required to activate the defects mentioned in Equation 4.7 is expressed as a function of separation of Si-H bonds and diffusion rate of H+ ion in the dielectric medium [38]. The applied or influenced electric field, application temperature, thickness of the oxide, and time are essentials parameters for modeling an accurate device lifetime. Another model was proposed for NBTI with the assumption that reaction-limited time dependence obeys a linear relationship. The subsequent reaction of hydrogen by-product affects the original reaction rate due to radiation. Hence a more refined model was proposed by M. A. Alam et al., which considers the generation of interface trap charges due to breaking of Si-H bonds, and the H+ ions generated by this reaction influence the process of diffusion dominantly [39].

Figure 4.3 illustrates the five steps of interface charge generation process with time. The behavior of curve is obtained by the following equations:

$$\frac{dN_{it}}{dt} = K_f \left[N_0 - N_{it} \right] - K_r N_{it} N_H^{(0)} \tag{4.9}$$

$$\frac{dNH}{dt} = DH \frac{d^2 NH}{dy^2} \tag{4.10}$$

In the above equations forward dissociation rate and reverse annealing rate of Si-H bond are denoted by K_f and K_r, respectively, and D_H is the diffusion constant. By modified boundary conditions of the first diffusion constant model we can easily generate these equations.

Step 1: N_{it} increases linearly due to the breaking of bond between silicon and hydrogen.

Step 2: Diffusion of hydrogen ion takes place; reaction is in equilibrium but the hydrogen flux away from the interface is negligible.

Step 3: Trap creation is limited by hydrogen diffusion (n = 1/4), which is independent of field and temperature.

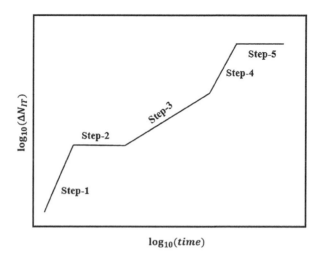

FIGURE 4.3   Interface charge generation process with time [28].

Step 4: Hydrogen diffuses into the gate with high diffusion velocity (n = 1/2).

Step 5: The maximum trap density, N_0 is reached due to the saturation process and N_{it} no longer increases with time (n = 0).

In some cases, trapping of charges can be possible by NBTI-like degradation, as the V_{th} shift of the device is caused by the tunneling of positively charged hot electrons injected from the gate metal [38–44]. The major factors that affect the NBTI can be process chemicals like H, N, Fl, and B; Si crystal orientation; temperature; and oxide growth process. To minimize these factors, certain methods can be considered, such as:

(1) Quality of oxide growth and creation of perfect Si/oxide and oxide/metal interface, which is nearly impossible, thus the selection of more resistant species for the chemical and electrical bond breaking.
(2) Focusing on stress and strain management, particularly at the interface, as the degradation of NBTI can be observed more frequently in strain silicon device structure.
(3) Use of deuterium (D_2) instead of hydrogen as much as possible due to its higher binding energy and also lower diffusing property than those of hydrogen.

4.5   USE OF HIGH-k DIELECTRIC

Before the use of high-k material, SiO_2, fabricated by the dry thermal oxidation, was used as the oxide layer in the MOS capacitor. Tunneling property of the charge particle through the dielectric layer was first found in 1965 [45]. The thin SiO_2 layers experience a tunnel current density of more than 1×10^{-9} A/cm^2, which was observed by many investigators [46–48]. The leakage currents caused by the trap charges can easily be calculated by the superposition principle, and thus we can build an idea of how the impurity present in the material can generate trap charges, which is shown Table 4.1.

As we discussed earlier in the section, SiO_2 possesses a lot of advantages; the only disadvantage is the leakage current due to the low dielectric constant. Since we are focusing here on the charge-trapping property of the dielectric layer, our objective is to compare different high-k materials with conventional SiO_2. Normally the high-i

TABLE 4.1

Trap Charge Property Near the Interface Region [49]

Element	Trap Energy > Si Valence Band Energy (eV)	Interface Area (cm²)
Mg	0.54	1.1×10^{-18}
Cr	0.20	5.2×10^{-17}
Cu	0.52	6.6×10^{-10}
Au	0.97	2.2×10^{-16}

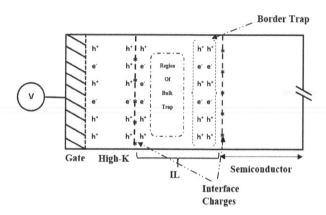

FIGURE 4.4   Trap charges at different interface in stack gate oxide structure.

dielectric carries up to $10^{13}/cm^2$ bulk and interface charge, and this value may go up when a gate-stacking technique is applied, whereas the Si–SiO$_2$ structure generally tops out at $10^{10}/cm^2$ of charge density. Due to the crystal mismatch while using high-*k*, use of an interfacial layer is necessary, which is called gate stacking (shown in Figure 4.4). This interfacial layer introduces another region between the IL and high-*k* to share more traps, which are more active than the bulk trap in the carrier exchange as discussed earlier.

Apart from the trap generation due to thermal and chemical fabrication steps, activation of unwanted ions in a dielectric layer is very minimal in case of SiO$_2$-S$_{poly}$ and S-SiO$_2$-Metal. Material with high dielectric constant also gives us some advantages over SiO$_2$, such as maximum reduction of lateral electric field across the gate stack due to higher permittivity ratio. This in turn helps us achieve channel inversion with much lower gate bias. So the main disturbance created by the high density of traps in the threshold voltage shift can be minimized, and the semiconductor node advancement in both constant voltage scaling and constant field scaling is also possible with high-*k* dielectric layers [50–52].

4.6   TRAP TIME CONSTANT

In the earlier brief discussion, we said that the necessity of interfacial layer (IL) is higher due to the crystal mismatch of the high-*k* layer and Si body. But it is also true that the use of IL can increase the overall trap density in a device. Figure 4.5 illustrates the energy band (EB) diagram of a P-MOS structure. The complete EB diagram is drawn with all the regions capable of trapping charges between the silicon layer and the gate metal. The time delay and surface potential (SP) function on the diagram-mentioned SP are considered as linear for ease of understanding. The most important parameter we need to focus on is the pseudo-Fermi level, which is given by the channel majority charge carrier for the section silicon-IL and IL-high-*k*, and the rest is governed by the gate Fermi level. For the purposes of application, the present defects are affected by applied bias and switching frequency of the device for the trapping and de-trapping of e− and /or h+. The factors responsible

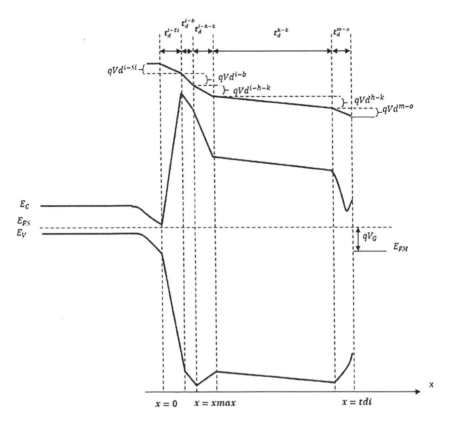

FIGURE 4.5 Energy band diagram with interfacial layer showing time constant for trap charges in different regions.

for the number of charge exchange are EOT and the potential barrier of the used dielectric. The chemical and electrical properties and energy level depth of the trapped charges may vary. The trapping and de-trapping time of charges reflect the nature of a particular device, except in the above case the calculation of this parameter is important when stacking dielectric layers used in a memory-based application. Theoretically, the insulator is a forbidden area for the movement of charge, so we can imagine how difficult it is to find the probability of the trapping and de-trapping of charge, and again the emission of e−/h+ is completely different from the capturing mechanism of the same. The trapping probability (TP) of electrons and holes can be calculated as the $V^e_{thermal} \times A^e_{cross\ section}$ and $V^h_{thermal} \times A^h_{cross\ section}$, respectively, and these cross-sections can vary from 10^{-12} to 10^{-18} cm² [53]. So we can say that the capturing time τ_t at any location is inversely proportional to the TP and the density of the charge at the point.

$$\left[\tau^e_t\right] P_x = 1 / \left\{\left(V^e_{thermal} \times A^e_{cross\ section}\right) \times D_e\, P_x\right\} \tag{4.11}$$

$$\left[\tau^h_t\right] P_x = 1 / \left\{\left(V^h_{thermal} \times A^h_{cross\ section}\right) \times D_h\, P_x\right\} \tag{4.12}$$

For our scaling perspective, the objective is to fabricate a dielectric layer with lower EOT. So for the thin layer of total oxide region, the interchanging of charge carrier takes place not only with the Si channel but also with the gate material. These exchanges can compete with each other for the trapping with the defects, so the areas with high integrity of free charge carrier and experienced external bias can manipulate device operations in lower EOT design.

4.7 PROPERTIES OF TRAP IN THE HIGH-*k* WITH INTERFACIAL LAYER

In this section we go through the nature of possible trapped charges in each of the available region and try to analyze those electrically. These charges can be characterized by the local allowed state, charge state, and exchanging cross-section. Characteristics of all the traps can only vary by the applied gate bias, i.e., if the charge state variations with the external bias are applied to the electric field, then it can affect the MOS capacitance.

4.7.1 SILICON AND INTERFACIAL LAYER

This is the region where transition happens from covalent semiconductor to covalent insulator. From the many experimental verifications on the Si–SiO$_2$ interface, we have reasons to believe that the maximum number of traps present in this region is due to the amphoteric material [54, 55], also known as the amphoteric trap. These traps can be charged negatively while having two electrons, and with one and no electron they carry neutral and positive change, respectively. In addition, some chemical-fabricated traps can also exist in this region [56, 57].

4.7.2 INTERFACIAL BULK REGION

With the silicon channel, the most effective IL used is either SiO$_2$ or SiON. These can be created by the dry thermal oxidation with or without nitrogen impurity, so the trap in the area is due to the oxygen or nitrogen vacancies with an energy band below the Si conduction band. Along with this, due to the stress factor at the IL and HK interface, some more stress-induced traps can be generated [58].

4.7.3 IL/HK INTERFACE

In this region we can find some of major and important trap density in the whole oxide layer due to the ionic and lattice mismatch. This vital region can also resist some of our most effective trap extraction techniques (discussed in the next section), such as MOS conductance and charge pumping. However, the width of the layer can also have an impact on the trapping and de-trapping of charges. The density of charge increases with the increase in distance from pure silicon [59, 60].

4.7.4 HK Bulk

The oxygen vacancies are the dominant factor for the charge trapping in the body of the HK layer. The O_2 vacancies exist in five different states from −2 to +2. The diffusion of O_2 vacancies occurs from the HK layer to the deficient interfacial layer, and this diffusion also depends on the process variable.

4.7.5 HK and Gate Interface

The conventional SiO_2 and metal (gate) interface is stable with almost zero trap density, whereas the HK interface is dominantly affected by the metal wave function, which creates a high density of trap charges. The major problem is that there is no technique available for the trap extraction from a metal region [61].

4.8 TECHNIQUE FOR THE TRAP EXTRACTION FROM THE STACK STRUCTURE

The earlier-mentioned techniques are less reliable for the trap charge extraction in a HK stack device due to its quasi-static characteristics. Due to complicated trapping and de-trapping methods of e−/h+ in the HK bulk and interfacial area, we need to consider some other important parameters with the available charge extraction methodology.

4.8.1 Capacitance Inversion Technique (CI)

With the gate oxide stack technique, the major problem arising in the trap extraction procedure from the c-v characteristics is the leakage from the gate side, due to IL trap density, which prevents the inversion layer to form at the Si channel by drawing some of the majority charge carrier, and hence creates a disturbance in the observation of CI curve. Some methods enhance the generation of minority charge carrier for better understanding. The technique involves dominating the number of charge carriers that participate in the generation and recombination of the trap at the IL or bulk dielectric layer. This is because of the photo illumination and temperature excitement which acts as the source of charge injection [62, 63].

4.8.2 Charge Pumping Technique

Unlike the charge inversion, charge pumping technique involves the non-steady-state procedure. In this method, a periodic pulse signal supplied at the gate operate the channel between accumulation and inversion modes, while the source and drain with substrate are excited by reverse bias. The charge pumping current (I_{cp}) measured from the substrate over the pulse duration carries the information of trapping and de-trapping of the charge carrier. The practical calculation of I_{cp} involves the calculation of cross-sectional area of exchange ($A^{e/h}_{cross\ section}$), trap density (D_{it}), and exchange location or point of charge exchange ($P_x^{e/h}$) independently [64, 65].

The most widely used HK material is Hafnium Dioxide (HfO_2) since the announcement by Intel in 2007 of its 45-node technology, and it is the first commercial production of FET with HK [66, 67]. The trapping property of HfO_2 can easily be investigated from the IV characteristics and the charge pumping as mentioned earlier in this chapter. With a positive gate bias, HfO_2 can be more capable in trapping of charge, especially electrons, in comparison to HfSiON, which can be clearly seen in its wider hysteresis curve [68].

As the fabrication or oxide growth procedure influence the trapping capability of dielectric layer, so different fabrication techniques highly influence the generation, trapping and de-trapping of charges. Atomic layer deposition (ALD) and physical vapor deposition for HfO_2 growth give almost similar results, whereas HfSiON can only be fabricated by the metal organic chemical vapor deposition (MOCVD). So the Hf defects present in the HfO_2 are the main reason for the trapping of electrons. Again, the interface trap at the gate electrode can be distinguishable in case of poly-Si gate due to the removal of the depletion layer using the metal gate electrode.

4.9 SUMMARY

The use of simple high-*k* or HK gate stack generates higher density of trap than the conventional SiO_2 dielectric layer. The interfacial layer with a high *k* comes with a benefit of crystal matching between Si channel and HK material, which maintains the required EOT for the particular application but also provides a larger region for trapping of charges and fabricated crystal defects. The electrical property of these traps usually differs from the SiO_2-based traps. So the inclusion of trap charges analysis while using HK with a multi-gate transistor is vital for the operation of device and its performance. Apart from that, for the memory-based application, charges-storing capability and efficiency with retention time highly depend on these charges. The circuits that are much more exposed to high radiation, like the ones used in space and biomedical applications, can produce irrational results with the presence of defects or traps.

REFERENCES

[1] T. Sekigawa, Y. Hayasi, "Calculated threshold voltage characteristics of an XMOS transistor having an additional bottom gate," *Solid State Electronics*, vol. 27, no. 8, pp. 827–828, 1984.

[2] G. Moore, "Cramming more component on to integrated circuit," *Electronics*, vol. 38, pp. 114, 1965.

[3] E. H. Nicollian, C. N. Berglund, P. F. Schmidt, and J. M. Andrew, "Electrochemical charging of thermal SiO_2 films by injected electron currents," *Journal of Applied Physics*, vol. 42, no. 13, pp. 5654–5664, 1971.

[4] D. K. Schroder and Jeff A. Babcock, "Negative bias temperature instability: road to cross in deep submicron silicon semiconductor manufacturing," *Journal of Applied Physics*, vol. 94, no. 1, pp. 1513–1530, 2003.

[5] B. E. Deal, M. Sklar, A. S. Grove, and E. H. Snow, "Characteristics of surface-state charge (QSS) of thermally oxidized silicon," *Journal of Electrochemical Society*, vol. 114, no. 3, pp. 266–274, 1967.

[6] C. E. Blat, E. H. Nicollian, and E. H. Poindexter, "Mechanism of negative-bias-temperature instability," *Applied Physics Letters*, vol. 69, no. 3, pp. 1712–1720, 1991.

[7] D. M. Fleetwood, X. J. Zhou, Leonidas Tsetseris, S. T. Pantelides, and R. D. Schrimpf. "Hydrogen model for negative-bias temperature instabilities in MOS gate insulators," *Silicon Nitride and Silicon Dioxide Thin Insulating Films and Other Emerging Dielectrics VIII*, pp. 267–278, 2005.

[8] J. A. Felix, M. R. Shaneyfelt, D. M. Fleetwood, T. L. Meisenheimer, J. R. Schwank, R. D. Schrimpf, P. E. Dodd, E. P. Gusev, and C. D'Emic. "Radiation-induced charge trapping in thin $Al_2O_3/SiO_xN_y/Si(100)$ gate dielectric stacks," *IEEE Transactions on Nuclear Science*, vol. 50, no. 6, pp. 1910–1918, 2003.

[9] J. A. Felix, D. M. Fleetwood, R. D. Schrimpf, J. G. Hong, G. Lucovsky, J. R. Schwank, and M. R. Shaneyfelt, "Total-dose radiation response of hafnium-silicate capacitors," *IEEE Transactions on Nuclear Science*, vol. 49, no. 6, pp. 3191–3196, 2002.

[10] M. Houssa, G. Pourtois, M. M. Heyns, and A. Stesmans, "Defect generation in high κ gate dielectric stacks under electrical stress: the impact of hydrogen," *Journal of Physics: Condensed Matter*, vol. 17, no. 21, pp. s2075–s2088, 2005.

[11] E. P. Gusev, E. Cartier, D. A. Buchanan, M. Gribelyuk, M. Copel, H. Okorn-Schmidt, and C. D'Emic, "Ultrathin high-κ metal oxides on silicon: processing, characterization and integration issues," *Microelectronic Engineering*, vol. 59, no. 4, pp. 341–349, 2001.

[12] G. D. Wilk, R. M. Wallace, and J. M. Anthony, "High-κ gate dielectrics: current status and materials properties considerations," *Applied Physics Letters*, vol. 89, no. 10, pp. 5243–5275, 2001.

[13] G. D. Wilk, R. M. Wallace, and J. M. Anthony, "Hafnium and zirconium silicates for advanced gate dielectrics," *Journal of Applied Physics*, vol. 87, no. 1, pp. 484–492, 2000.

[14] E. P. Gusev, D. A. Buchanan, E. Cartier, A. Kumar, et al. "Ultrathin high-κ gate stacks for advanced CMOS devices," *IEEE International Electron Devices Meeting (IEDM) Technical Digest*, pp. 451–454, 2001.

[15] G. D. Wilk, and R. M. Wallace, "Electrical properties of hafnium silicate gate dielectrics deposited directly on silicon," *Applied Physics Letters*, vol. 74, no. 19, pp. 2854–2856, 1999.

[16] L. Kang, B. H. Lee, W. J. Qi, Y. Jeon, R. Nieh, S. Gopalan, K. Onishi, and J. C. Lee, "Electrical characteristics of highly reliable ultrathin hafnium oxide gate dielectric," *IEEE Electron Device Letters*, vol. 21, no. 4, pp. 181–183, 2000.

[17] B. H. Lee, L. Kang, W. J. Qi, R. Nieh, Y. Jeon, K. Onishi, and J. C. Lee, "Ultrathin hafnium oxide with low leakage and excellent reliability for alternative gate dielectric application," *IEEE International Electron Devices Meeting (IEDM) Technical Digest*, pp. 133–135, 1999.

[18] G. Lucovsky and G. B. Rayner Jr., "Microscopic model for enhanced dielectric constants in low concentration SiO_2-rich noncrystalline Zr and Hf silicate alloys," *Applied Physics Letters*, vol. 77, no. 18, pp. 2912–2914, 2000.

[19] S. M. Sze, *Physics of Semiconductor Devices*, New York: John Wiley & Sons, 1981.

[20] E. H. Nicollian and J. R. Brews, *MOS (Metal Oxide Semiconductor) Physics and Technology*, New York: John Wiley & Sons, 1982.

[21] F. B. McLean and T. R. Oldham, "Basic mechanisms of radiation effects in electronic materials and devices," *Harry Diamond Laboratories Technical Report*, no. HDL-TR-2129, September 1987.

[22] J. A. Felix, J. R. Schwank, D. M. Fleetwood, M. R. Shaneyfelt, and E. P. Gusev, "Effects of radiation and charge trapping on the reliability of high-κ gate dielectrics," *Microelectronics Reliability*, vol. 44, no. 4, pp. 563–575, 2004.

[23] J. A. Felix, H. D. Xiong, D. M. Fleetwood, E. P. Gusev, R. D. Schrimpf, A. L. Sternberg, and C. D'Emic, "Interface trapping properties of nMOSFETs with $Al_2O_3/SiO_xN_y/$ Si(100) gate dielectric stacks after exposure to ionizing radiation," *Microelectronic Engineering*, vol. 72, no. 1–4, pp. 50–54, 2004.

[24] P. M. Lenahan and P. V. Dressendorfer, "Hole traps and trivalent silicon centers in metal/ oxide/silicon devices," *Journal of Applied Physics*, vol. 55, no. 10, pp. 3495–3499, 1984.

[25] T. P. Ma, G. Scoggan, and R. Leone, "Comparison of interface-state generation by 25-keV electron beam irradiation in p-type and n-type MOS capacitors," *Applied Physics Letters*, vol. 27, no. 2, pp. 61–63, 1975.

[26] P. S. Winokur, J. R. Schwank, P. J. McWhorter, P. V. Dressendorfer, et al. "Correlating the radiation response of MOS capacitors and transistors," *IEEE Transactions on Nuclear Science*, vol. 31, no. 6, pp. 1453–1460, 1984.

[27] D. M. Fleetwood, "Border traps in MOS devices," *IEEE Transactions on Nuclear Science*, vol. 39, no. 2, pp. 269–271, 1992.

[28] D. M. Fleetwood, W. L. Warren, J. R. Schwank, P. S. Winokur, M. R. Shaneyfelt, and L. C. Riewe, "Effects of interface traps and border traps on MOS postirradiation annealing response," *IEEE Transactions on Nuclear Science*, vol. 42, no. 6, pp. 1698–1707, 1995.

[29] R. E. Stahlbush, A. H. Edwards, D. L. Griscom, and B. J. Mrstik, "Post-irradiation cracking of H_2 and formation of interface states in irradiated metal-oxide-semiconductor field-effect transistors," *Journal of Applied Physics*, vol. 73, no. 2, pp. 658–667, 1993.

[30] R. E. Stahlbush, E. Cartier, and D. A. Buchanan, "Anomalous positive charge formation by atomic hydrogen exposure," *Microelectronic Engineering*, vol. 28, no. 1–4, pp. 15–18, 1995.

[31] D. J. DiMaria, "Temperature dependence of trap creation in silicon dioxide," *Journal of Applied Physics*, vol. 68, no. 10, pp. 5234–5246, 1990.

[32] D. M. Fleetwood, M. R. Shaneyfelt, W. L. Warren, J. R. Schwank, T. L. Meisenheimer, and P. S. Winokur, "Border traps: issues for MOS radiation response and long-term reliability," *Microelectronics Reliability*, vol. 35, no. 3, pp. 403–428, 1995.

[33] D. M. Fleetwood, M. P. Rodgers, L. Tsetseris, X. J. Zhou, et al. "Effects of device aging on microelectronics radiation response and reliability," *Microelectronics Reliability*, vol. 47, no. 7, pp. 1075–1085, 2007.

[34] E. H. Nicollian, C. N. Berglund, P. F. Schmidt, and J. M. Andrew, "Electrochemical charging of thermal SiO_2 films by injected electron currents," *Journal of Applied Physics*, vol. 42, no. 13, pp. 5654–5664, 1971.

[35] L. Tsetseris, X. J. Zhou, D. M. Fleetwood, R. D. Schrimpf, and S. T. Pantelides, "Field-induced reactions of water molecules at Si-dielectric interfaces," *Materials Research Society Symposium Proceedings*, vol. 786, pp. 171–176, Boston, 2004.

[36] D. M. Fleetwood, M. P. Rodgers, L. Tsetseris, X. J. Zhou, I. Batyrev, S. Wang, R. D. Schrimpf, and S. T. Pantelides, "Effects of device aging on microelectronics radiation response and reliability," *Microelectronics Reliability*, vol. 47, no. 7, pp. 1075–1085, 2007.

[37] V. Huard, M. Denais, F. Perrier, N. Revil, C. Parthasarathy, A. Bravaix, and E. Vincent, "A thorough investigation of MOSFETs NBTI degradation," *Microelectronics Reliability*, vol. 45, no. 1, pp. 83–98, 2005.

[38] K. O. Jeppson, and C. M. Svensson, "Negative bias stress of MOS devices at high electric fields and degradation of MNOS devices," *Journal of Applied Physics*, vol. 48, no. 5, pp. 2004–2014, 1977.

[39] M. A. Alam and S. Mahapatra, "A comprehensive model of PMOS NBTI degradation," *Microelectronics Reliability*, vol. 45, no. 1, pp. 71–81, 2005.

[40] M. Houssa, "Modelling negative bias temperature instabilities in advanced p-MOS-FETs," *Microelectronics Reliability*, vol. 45, no. 1, pp. 3–12, 2005.

[41] S. Tsujikawa and J. Yugami, "Positive charge generation due to species of hydrogen during NBTI phenomenon in pMOSFETs with ultra-thin SiON gate dielectrics," *Microelectronics Reliability*, vol. 45, no. 1, pp. 65–69, 2005.

[42] S. Mahapatra and M. A. Alam, "A predictive reliability model for PMOS bias temperature degradation," *IEEE International Electron Devices Meeting (IEDM) Digest*, pp. 505–508, December 2002.

[43] Y. Hiruta, H. Iwai, F. Matsuoka, K. Hama, et al. "Interface state generation under long term positive bias temperature stress for a p+ poly gate MOS structure," *IEEE Transactions on Electron Devices*, vol. 36, no. 9, pp. 1732–1739, 1989.

[44] S. Maeda, et al., "Mechanism of negative-bias temperature instability in polycrystalline-silicon thin film transistors," *Journal Applied Physics*, vol. 76, no. 12, pp. 8160–8166, 1994.

[45] H. C. Card and E. H. Rhoderick. "Studies of tunnel MOS diodes I. Interface effects in silicon Schottky diodes," *Journal of Physics D: Applied Physics*, vol. 4, no. 10, pp. 1589, 1971.

[46] H. C. Card and E. H. Rhoderick, "Studies of tunnel MOS diodes II. Thermal equilibrium considerations," *Journal of Physics D: Applied Physics*, vol. 4, no. 10, pp. 1602–1611, 1971.

[47] M. A. Green and J. Shewchun, "Current multiplication in metal-insulator-semiconductor (MIS) tunnel diodes," *Solid-State Electron*, vol. 17, no. 4, pp. 349–365, 1974.

[48] M.A. Green and J. Shewchun, "Application of the small-signal transmission line equivalent circuit model to the ac, dc and transient analysis of semiconductor devices," *Solid-State Electronics*, vol. 17, no. 9, pp. 941–949, 1974.

[49] S. Kar and W. E. Dahlke, "Interface states in MOS structures with 20-40 a thick SiO2 films on nondegenerate Si," *Solid-State Electron*, vol. 15, no. 2, pp. 221–237, 1972.

[50] S. Kar, "Nature of Interface Traps in Si/SiO$_2$/HfO$_2$/TiN Gate Stacks and its Correlation with the Flat-band Voltage Roll-off," *ECS Transactions*, vol. 25, no. 6, pp. 399, 2009.

[51] R. Choi, S. J. Rhee, J. C. Lee, B. H. Lee, et al. "Charge trapping and detrapping characteristics in hafnium silicate gate stack under static and dynamic stress," *IEEE Electron Device Letters*, vol. 26, no. 3, pp. 197–199, 2005.

[52] G. Bersuker, J. H. Sim, C. S. Park, C. D. Young, et al. "Mechanism of electron trapping and characteristics of traps in HfO$_2$ gate stacks," *IEEE Transactions on Device and Materials Reliability*, vol. 7, no. 1, pp. 138–145, 2007.

[53] A. Rose, "Concepts in photoconductivity and allied problems," No. 19. Interscience publishers, 1963.

[54] G. J. Gerardi, E. H. Poindexter, P. J. Caplan, and N. M. Johnson, "Interface traps and Pb centers in oxidized (100) silicon wafers," *Applied Physics Letter*, vol. 49, no. 6, pp. 348–350, 1986.

[55] A. H. Edwards, "Theory of the Pb center at the <111> Si/SiO$_2$ interface," *Physical Review B*, vol. 36, no. 18, pp. 9638, 1987.

[56] J. Dong and D. A. Drabold, "Atomistic structure of band-tail states in amorphous silicon," *Physical Review Letter*, vol. 80, no. 9, pp. 1928–1931, 1998.

[57] S. Kar and W. E. Dahlke, "Interface states in MOS structures with 20–40 Å thick SiO$_2$ films on nondegenerate Si." *Solid-State Electronics*, vol. 15, no. 2, pp. 221–237, 1972.

[58] A. V. Kimmel, P. V. Sushko, A. L. Shluger, and G. Bersuker, "Positive and negative oxygen vacancies in amorphous silica," *ECS Transactions*, vol. 19, no. 2, p. 3, 2009.

[59] A. Toriumi and K. Kita, "On the origin of anomalous V$_{TH}$ shift in high-k MOSFETs," *ECS Transactions*, vol. 19, no. 1, pp. 243, 2009.

[60] H. Jagannathan, V. Narayanan, and S. Brown, "Engineering high dielectric constant materials for band-edge CMOS applications," *ECS Transactions*, vol. 16, no. 5, pp. 19, 2008.

[61] J. Tersoff, "Schottky barrier heights and the continuum of gap states," *Physical Review Letters*, vol. 52, no. 6, pp. 465, 1984.

[62] T. C. Poon and H. C. Card, "Energy and electric field dependence of Si-SiO$_2$ interface state parameters by optically activated admittance experiments," *Journal of Applied Physics*, vol. 51, no. 12, pp. 6273–6278, 1980.

[63] S. Kar and S. Varma, "Determination of silicon-silicon dioxide interface state properties from admittance measurements under illumination," *Journal of Applied Physics*, vol. 58, no. 11, pp. 4256–4266, 1985.

[64] J. S. Brugler and P. G. Jespers, "Charge pumping in MOS devices," *IEEE Transactions on Electron Devices*, vol. 16, no. 3, pp. 297–302, 1969.

[65] G. Groeseneken, H. E. Maes, N. Beltran, and R. F. de Keersmaecker, "A reliable approach to charge-pumping measurements in MOS transistors," *IEEE Transactions on Electron Devices*, vol. 31, no. 1, pp. 42–53, 1984.

[66] A. Nishiyama, "Hafnium-based gate dielectric materials," *In High Permittivity Gate Dielectric Materials*, Springer, pp. 153–181, Berlin, Heidelberg, 2013.

[67] G. D. Wilk, R. M. Wallace, and J. M. Anthony, "High-κ gate dielectrics: Current status and materials properties considerations," *Journal of Applied Physics*, vol. 89, no. 10, pp. 5243–5275, 2001.

[68] M. T. Bohr, R. S. Chau, T. Ghani, and K. Mistry, "The high-k solution," *IEEE Spectrum*, vol. 44, no. 10, pp. 29–35, 2007.

5 Impact of High-*k* Dielectric on the Gate-Induced Drain Leakage of Multi-Gate FETs

Varshini K. Amirtha
National Institute of Technology, Tiruchirappalli, India

Shubham Sahay
Indian Institute of Technology, Kanpur, India

CONTENTS

5.1 INTRODUCTION

Evolution is inevitable and mandatory in all sectors of this expeditiously growing world. There has been a tremendous change in the landscape of consumer electronics in the past few decades. For instance, giant room-sized computers have been replaced

DOI: 10.1201/9781003121589-5

by handheld smart devices that have the processing capability of a supercomputer. The significant improvement in semiconductor technology has driven this evolution of electronic devices. In this era of big data, internet-of-things (IoT) devices such as laptops, mobile phones, smart watches, tablets, wireless sensor nodes, smart home appliances, etc., which form a part of the cyber-physical system, have become an integral part of our daily life. The semiconductor integrated circuits (ICs) embedded in these smart devices consist of billions of transistors. The semiconductor industry has thrived because of the incessant scaling of these transistors to reduce the area and energy consumption of these ICs and integrate several ICs on a single device to realize multiple functionalities. The conventional planar metal-oxide-semiconductor field-effect transistors (MOSFETs) continued to dominate the semiconductor industry until the early 2000s. However, scaled MOSFETs exhibit severe short channel effects (SCEs), which significantly increases their power consumption and hinders their relentless scaling. Therefore, architectures with enhanced gate control, i.e., gate electrodes controlling the channel region from more than one side, popularly known as multi-gate FETs, such as FinFETs, were subsequently proposed beyond the 22-nm technology node to sustain the CMOS scaling process [1]. Although FinFETs continued to be the prominent choice of the semiconductor tech giants until the 7-nm technology node, advanced architectures where the gate wraps around the channel from all sides such as gate-all-around nanowire (GAA-NW) FETs, nanotube (NT) FETs, and stacked nano-sheet (NS) FETs are being explored for scaling the CMOS technology to the 5-nm technology node and beyond. The GAANWFETs, NTFETs and NSFETs exhibit a significantly improved electrostatic control of the channel region as compared to FinFETs, leading to an enhanced immunity to SCEs. However, the efficient modulation of the channel region results in an undesired overlap of the conduction band of the drain region with the valence band of the channel region in the OFF state, leading to a new lateral band-to-band tunneling (L-BTBT) leakage current. Since this tunneling phenomenon arises due to the improved gate control, it is categorized as a component of gate-induced drain leakage (GIDL).

Moreover, insulators other than SiO_2, such as HfO_2, Si_3N_4, Al_2O_3, etc., which have a higher dielectric constant (k), have also been explored as gate oxide and gate-sidewall spacer material to optimize the performance of the MuGFETs. The application of a high-k dielectric facilitates the realization of an effective oxide thickness (EOT), which is smaller than the physical thickness of the high-k dielectric. The EOT indicates the equivalent thickness of SiO_2 required to achieve the same electrostatics which is provided by a high-k dielectric of a particular physical thickness as:

$$EOT = T_{physical,high-k}\left(\frac{\varepsilon_{Si}}{\varepsilon_{high-k}}\right)$$

where $T_{physical,high-k}$ is the physical thickness of the high-k dielectric, ε_{Si} and ε_{high-k} are the relative permittivity of SiO_2 and high-k dielectric, respectively.

The high-k dielectric allows the utilization of a gate oxide with a higher physical thickness while achieving a low EOT and significantly reduces the possibility of direct tunneling–based gate leakage current, which originates due to the scaling of the gate oxide (SiO_2). Also, the application of a high-*k* dielectric as gate side-wall spacer increases the effective channel length due to increased fringing field and boosts the immunity of MuGFETs against SCEs.

Since the presence of a high-*k* dielectric in the gate oxide or gate-sidewall spacer significantly modulates the electrostatics of a MuGFET, it may also be effective in the optimization of the L-BTBT GIDL component in the emerging MuGFETs. To this end, this chapter provides a comprehensive analysis of the impact of high-*k* dielectric in the gate oxide and gate-sidewall spacer on the L-BTBT GIDL characteristics of emerging MuGFETs such as NWFETs, NTFETs, and NSFETs. First, let us discuss the L-BTBT GIDL and its implications on the emerging MuGFETs in detail.

5.2 GATE-INDUCED DRAIN LEAKAGE (GIDL)

The major challenge while designing semiconductor devices is to increase the speed while reducing the power consumption and area. VLSI designers were able to satisfy this stringent constraint through CMOS scaling process following the Moore's law. The reduction in the length of the MOSFETs not only increases the current (and, hence, speed) but also leads to a reduction in the area and capacitance which manifests itself as a reduction in the dynamic power dissipation [2]. However, the incessant scaling of MOSFETs also leads to several short channel effects such as threshold voltage roll-off, drain-induced barrier lowering, hot carrier injection, etc., which significantly increase the static leakage current of the MOSFETs [3]-[4]. As a result, the static power dissipation dominates over the dynamic power dissipation as the Moore's law approaches its limits. Another implication of the CMOS scaling process is the gate-induced drain leakage (GIDL) phenomenon. Owing to the scaling of the gate-oxide thickness along with the channel length, the vertical electric field along the gate-oxide thickness increases significantly. This electric field depletes the drain region under the gate electrode. GIDL originates due to the band-to-band tunneling (BTBT) of electrons within the drain region under the gate–drain overlap area. In the conventional MOSFETs, this transverse (T) BTBT constitutes the GIDL current and is dominant at high drain voltages and low gate voltages. In the next section, we shall briefly look at the components of GIDL in MuGFETs.

5.2.1 TRANSVERSE-BTBT

The conventional planar MOSFETs are fabricated using the self-aligned gate technology. In this process, the inherent diffusion of dopant atoms from the source and drain regions toward the channel region under the gate electrode leads to a gate–drain overlap (Figure 5.1a). When a positive drain voltage and a negative gate voltage are applied to the MOSFETs with ultralow gate-oxide thickness, the drain region under the gate electrode is depleted. This results in a large band bending in the drain region and alignment of the valence band of the drain region near the surface with the conduction band of the drain region far away from the surface, leading to tunneling of electrons within the drain region, as shown in Figure 5.1b. This transverse tunneling (T-BTBT) leads to a high leakage current and degrades the performance of

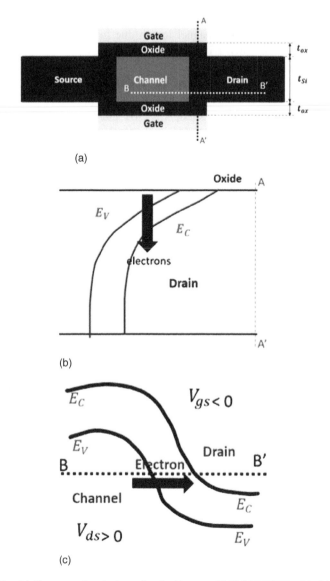

FIGURE 5.1 (a) Cross-sectional view of a double-gate (DG) MOSFET with (b) transverse BTBT component of GIDL along the cut-line A-A' and (c) lateral BTBT component of GIDL along cut-line B-B'.

the planar MOSFETs. The significant increase in the drain current of conventional MOSFETs at negative gate voltages is attributed to this T-BTBT [5–9]. Since T-BTBT depends on the fabrication process and doping profile of the MOSFETs, it can be reduced by using a lightly doped drain region or advanced doping processes with less dopant diffusion such as in situ doping, which limit the possibility of gate–drain overlap [10, 11].

5.2.2   Lateral-BTBT GIDL

The gate–drain overlap is inevitable in the conventional MOSFETs owing to the inherent doping process. On the other hand, the emerging MuGFETs such as NWFETs, NTFETs, etc. are fabricated using advanced manufacturing processes, including epitaxial growth and in situ doping techniques that restrict the gate–drain overlap. Therefore, T-BTBT should be significantly suppressed in the emerging MuGFETs. However, the experimental characteristics of the emerging MuGFETs indicate a large leakage current when the gate voltage is low and negative. This observation could only be explained with the introduction of a lateral (L) BTBT component of GIDL. The improved gate control of the channel region in the emerging MuGFETs not only reduces the short channel effects but also leads to an alignment of the channel region's valence band with the drain region's conduction band in the OFF state. This overlap facilitates lateral tunneling of electrons from the channel region to the drain region, as shown in Figure 5.1c. Therefore, unlike T-BTBT, which occurs when the gate voltage is negative, the L-BTBT is dominant even when the MuGFETs are in the OFF state and degrades their energy efficiency considerably. Several architectures have been proposed to minimize the impact of L-BTBT GIDL, such as dual-material gate, core-shell structure with various doping schemes, etc.

5.3   IMPACT OF HIGH-k DIELECTRIC ON GIDL OF MUGFETS

With this brief introduction to the L-BTBT GIDL phenomena, let us look at the different emerging MuGFETs in detail. Depending on the inherent architecture of the MuGFETs, the L-BTBT GIDL manifests itself in different forms and degrades their performance significantly. We shall also study how the incorporation of high-k dielectric in the gate oxide and gate side-wall spacer impacts the performance of the emerging MuGFETs in the subsequent sections.

5.3.1   Nanowire FETs

The FinFETs exhibit undesirable variability and high leakage when scaled beyond the 5-nm technology node [12]. Nanowire (NW) FETs have been proposed as a replacement for FinFET technology. In NWFETs, the gate electrode is wrapped around the channel and provides excellent electrostatic control [13–16]. The NWFETs exhibit significantly large ON-current to OFF-current ratio (I_{ON}/I_{OFF}), an improved sub-threshold slope close to the Boltzmann limit of 60 mV/decade, and small leakage current, making them attractive for logic applications.

5.3.1.1   Structure

The 3D schematic of a NWFET is shown in the Figure 5.2. Here, we can see a cylindrical silicon body with circular cross-section. However, NWFETs may also be fabricated to have a rectangular or trapezoidal cross-section. The gate electrode and the gate oxide layer are wrapped around the silicon body [18,19]. The efficient electrostatic control and improved sub-threshold slope in the NWFETs as compared with FinFETs are attributed to this gate-all-around (GAA) structure [20,21]. NWFETs

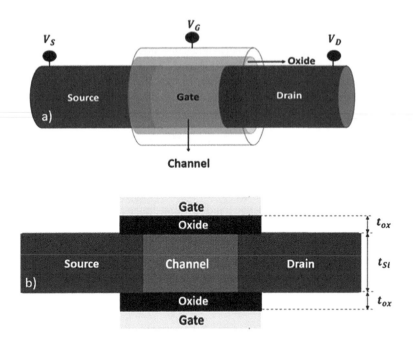

FIGURE 5.2 (a) Three-dimensional view and (b) cross-sectional view of GAA NWFET.

may be fabricated via top-down approaches based on advanced lithography techniques and etching methods [22]. However, the NWFETs fabricated using bottom-up techniques such as the vapor-solid-liquid (VSL) growth technique exhibit a higher quality as compared to those realized using the top-down approach.

5.3.1.2 GIDL in NWFETs

Since the NWFETs are fabricated using advanced manufacturing processes, including epitaxial growth and in situ doping techniques that restrict the gate–drain overlap, NWFETs are expected to have significantly suppressed GIDL. However, it has been observed experimentally that the NWFETs exhibit a high OFF-current, and the drain current increases when a negative gate voltage is applied, even with a fixed gate–drain potential V_{GD}. This clearly indicates that the T-BTBT GIDL is not responsible for the increase in the drain current with negative gate voltages. A detailed investigation reveals that there is a significant band overlap between the conduction band of the drain region and the conduction band of the channel region, owing to the strong gate coupling, which leads to a lateral BTBT (L-BTBT) at the channel–drain junction [23–28]. The electrons tunnel from the channel to the drain, unlike the conventional T-BTBT GIDL, where vertical tunneling takes place in the gate–drain overlap region within the drain [30,31]. Since L-BTBT is dominant even in the OFF state, it governs the OFF-state leakage current of the NWFETs and dictates their static power dissipation.

    Also, junctionless (JL) FETs, which do not contain any metallurgical junction, were proposed to alleviate the stringent requirement of an ultra-steep doping profile of MOSFETs. Although NWJLFETs exhibit a significantly suppressed L-BTBT as compared to NWMOSFETs owing to the relatively lightly doped drain region, they exhibit a low ON-current due to the increased series resistance offered by the lightly doped source/drain regions. Therefore, junctionless accumulation mode (JAM) FETs were proposed to increase the ON-current by replacing the lightly doped source and drain regions in JLFETs with heavily doped source/drain regions. Although the NWJAMFETs offer a higher ON-current as compared to the NWJLFETs, their L-BTBT GIDL characteristics are similar to the MOSFETs [32, 33].

5.3.1.3 Impact of High-*k* Sidewall Spacers

Gate-sidewall spacers are required for isolating the gate electrode with the source/drain electrodes and are inevitable while fabricating NWFETs. The source/drain region under the gate-sidewall spacers is known as the source/drain extension region. The addition of high-*k* spacers over the source/drain extensions in the conventional NWMOSFETs leads to suppression in the OFF current, as shown in Figure 5.3. The presence of a high-*k* spacer reduces the peak electric field at the channel–drain metallurgical junction, as shown in Figure 5.4a, and the fringing fields augment the electric field intensity in the drain extension region. The combined effect results in an increased tunneling width and a diminished L-BTBT, as shown in Figure 5.4b [17]. Hence, with the incorporation of high-*k* spacers, the L-BTBT GIDL-induced OFF-current reduces in conventional NWMOSFETs.

Gate–drain underlap is a useful technique to mitigate the T-BTBT GIDL current in conventional MOSFETs. Besides, the transfer characteristics of the underlapped NWFETs with different gate oxide materials are compared with the conventional NWFET with air spacer, as shown in Figure 5.5a. It can be observed from Figure 5.5a that the OFF-current is negligible in underlapped NWMOSFETs as compared with the conventional NWMOSFETs. This clearly indicates that the underlap architecture is also effective in dissuading the L-BTBT component of GIDL. However, the increased series resistance of the lightly doped channel region in the underlapped NWMOSFETs leads to reduction in the ON-current as compared with the conventional NWMOSFETs. The ON-current may be improved in the underlapped NWMOSFETs by including high-*k* spacers.

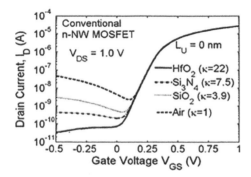

FIGURE 5.3 Transfer characteristics of NWMOSFETs with different gate sidewall-spacers [17].

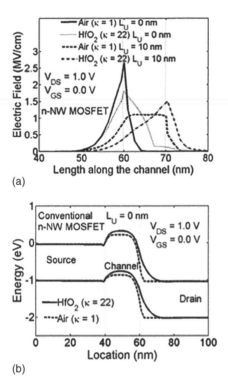

FIGURE 5.4 (a) Electric field of the conventional NWMOSFET and underlapped NWMOSFET with air and HfO₂ spacers and (b) energy band profiles of the conventional NWMOSFET with HfO₂ and air spacer [17].

However, the L-BTBT also increases in the underlapped NWMOSFETs in the presence of high-*k* spacers due to an increase in the electric field at the channel–drain interface owing to the fringing fields. The enhanced electric field leads to smaller tunneling width, increasing the OFF-state current (Figure 5.5b). Also, the total gate capacitances due to the higher fringing fields degrade the dynamic performance of the NWMOSFETs with and without underlap owing to the high-*k* spacer, as shown in Figure 5.6. Therefore, the usage of high-*k* dielectric in NWFETs provides an additional knob for optimizing their static and dynamic performance and reducing their L-BTBT GIDL-induced leakage current [34–38].

The OFF-current significantly increases when a high-*k* dielectric is utilized. This is attributed to the enlarged vertical electric field due to the presence of high-*k* dielectric, which increases the band overlap and facilitates L-BTBT GIDL.

5.3.2 NANOTUBE FET

Nanotube (NT) FETs can be described as NWFETs with an additional core gate. The presence of the core gate enhances the gate controllability of the channel region.

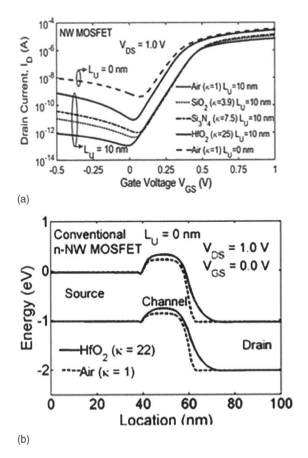

(a)

(b)

FIGURE 5.5   (a) Transfer characteristics of the conventional NWMOSFET with air spacer and the underlapped NWMOSFET with different gate sidewall spacers and (b) energy band profiles of the conventional NWMOSFET with air spacer and underlapped NWMOSFET with HfO_2 and air spacer [17].

Therefore, NTFETs exhibit a better immunity to the short channel effects and enhanced drive current capability as compared to NWFETs, making them an effective replacement for the NWFETs.

5.3.2.1 Structure

The 3D view of a NTFET is shown in Figure 5.7. Unlike the NWFETs with a cylindrical body, NTFETs have a cylindrical tube structure. The cylindrical silicon tube is wrapped by a shell gate from the outside and also filled with an inner core gate [40–44]. The diameter of the core gate is minimized in order to lower the Miller capacitance [45]. The NTFETs may be realized experimentally using the process flow described in [46,47] using epitaxial growth and in situ doping processes.

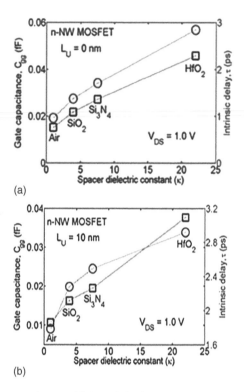

(a)

(b)

FIGURE 5.6 Gate capacitance (C_{gg}) and intrinsic delay (τ) of (a) the conventional NWMOSFET and (b) the underlapped NWMOSFET [17].

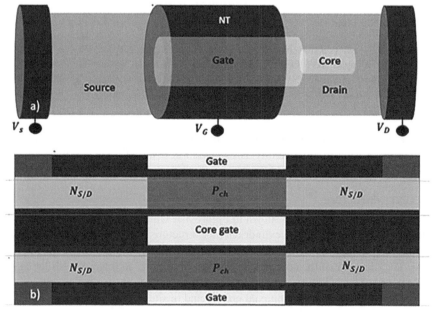

FIGURE 5.7 (a) Three-dimensional view and (b) cross-sectional view of the nanotube MOSFETs.

5.3.2.2   GIDL in Nanotube FET

Although NTFETs are considered superior to the NWFETs in terms of gate controllability, the L-BTBT GIDL arises due to the efficient gate control. Therefore, the
L-BTBT GIDL characteristics of NTFETs must be critically evaluated before pro-
jecting it as a successor to the NWFETs. Figure 5.9a shows the transfer characteristics of the NTFETs and NWFETs. It can be observed that NTFETs not only have an
increased ON-current but also exhibit a higher OFF-current. This results in a lower
I_{ON}/I_{OFF} in NTFET as compared to NWFETs.

In the NWFETs, the L-BTBT arises only due to the outer gate, whereas in
NTFETs, the core gate also contributes to the L-BTBT GIDL, which increases the
leakage current. The band alignment necessary for L-BTBT is facilitated at the channel–drain interface close to the Si–SiO$_2$ interface near the core gate as well as the
shell gate in NTFETs. Besides, the impact of gate scaling is investigated for the
NWFETs and NTFETs. It is clear from Figure 5.8a that the OFF-state current
increases considerably for the NTFETs compared with NWFETs owing to the additional L-BTBT component. However, the dynamic performance is improved in
NTFETs as compared to the NWFETs owing to the higher current driving capability
of NTFETs. It has also been observed that the gate scaling leads to drastic degradation in the performance of NTFETs as compared with NWFETs, which can be
observed in Figure 5.8b, [39].

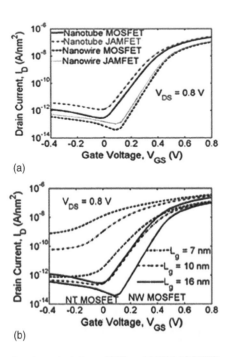

(a)

(b)

FIGURE 5.8   (a) Transfer characteristics of NT and NW MOSFETs and JAMFETs, and (b)
impact of gate length scaling on NT and NW MOSFETs [39].

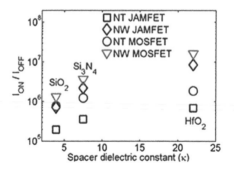

FIGURE 5.9   Transfer characteristics of NTFETs and NWFETs with different high-*k* gate spacers [39].

5.3.2.3   Impact of High-*k* Sidewall Spacers

The presence of a high-*k* dielectric in the gate-oxide significantly increases the L-BTBT GIDL due to the increased electric field. Therefore, low-*k* dielectrics should be preferred while designing NTFETs from a L-BTBT GIDL perspective [48, 49]. Moreover, the spacers are indispensable while fabricating the NTFETs. Figure 5.9 compares the transfer characteristics of the NTFETs with different spacer materials. As can be observed from Figure 5.10, the presence of a high-*k* spacer leads to a significant reduction in the NTFETs. The fringing fields lead to an attenuation in the band transition and a consequent increase in the tunneling width in the NTFETs. This results in a suppressed L-BTBT GIDL in NTFETs with the high-*k* spacer. However, the fringing fields increase the parasitic capacitances, resulting in a degradation of the dynamic performance [50, 51].

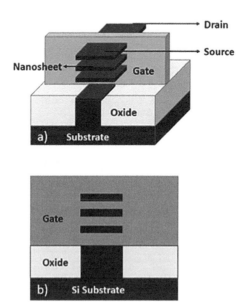

FIGURE 5.10   (a) Three-dimensional view and (b) cross-sectional view of nanosheet (NS) FETs.

5.3.3 Nano-Sheet FETs

Conventional FET architectures such as FinFETs do not allow much flexibility, and the fin heights cannot be increased further without interfering with the interconnect layer, making them unsuitable for the future technology nodes [52–58]. Therefore, NWFETs were proposed for the next generation of CMOS devices. However, the lateral NWFETs exhibit a low ON-current owing to the limited nanowire width for current flow [59–61]. Therefore, nano-sheet (NS) FETs that offer a higher active area for current flow were proposed. NSFETs are hailed as a promising candidate for the future CMOS technology nodes.

5.3.3.1 Structure

The schematic view of the NSFET architecture is shown in Figure 5.11. In NSFETs, several stacks of nano-sheets are bridged between the source and drain regions. Similar to NWFETs, the gate completely surrounds the channel region and provides a better electrostatic control. The usage of thin sheets of silicon in NSFETs increases the gate control while maintaining a higher effective area for current flow [62, 63]. This structure also provides flexibility, which helps in altering the sheet width. The sheet size can be tuned to boost the current or to reduce the power consumption. Nano-sheets may be fabricated using advanced atomic layer deposition processes. Apart from Si and SiGe, materials like Ge and InGaAs are also being explored to develop high-mobility NSFETs [64, 65].

5.3.3.2 GIDL in Nano-Sheet FET

Due to their gate-all-around architecture, even NSFETs exhibit a high immunity against the SCEs. However, the enhanced gate controllability leads to a band alignment between the conduction band of the drain junction and the valence band of the channel region, and hence facilitate L-BTBT. As a consequence, their OFF-current and static-leakage power increases, degrading their performance [66].

5.3.3.3 Impact of High-*k* Sidewall Spacers

In complex architectures such as stacked nano-sheet FETs, there is a limited scope for introducing structural innovations that are conventionally used to diminish the

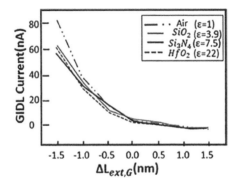

FIGURE 5.11   GIDL current of nanosheet FETs with different high-*k* gate sidewall spacers.

impact of L-BTBT GIDL. However, the gate-sidewall spacer is inevitable even for NSFETs, and spacer optimization is reported as a promising technique to reduce the L-BTBT GIDL [67–69]. From Figure 5.11, it can be observed that the spacer material influences the GIDL in NSFETs. The fringing field emanating out of the high-*k* spacers introduce a dispersion in the energy band profiles and leads to a larger tunneling width. This reduces the L-BTBT and the GIDL current [70]. However, the overall parasitic capacitance also increases with an increase in the permittivity of the high-*k* spacer [71, 72]. The use of a low-*k* spacer reduces the parasitic capacitance, but it also suppresses the ON-current [73]. Therefore, to obtain a low GIDL current along with an improved dynamic performance, dual-*k* spacers with a combination of high-*k* and low-*k* spacers were proposed [74, 75]. NSFETs with dual-*k* spacers are potential candidates for achieving low OFF-current, high ON-current, and improved dynamic performance [76].

5.4   CONCLUSION

In this chapter, we have provided a comprehensive analysis of the impact of high-*k* dielectric on the behavior of emerging MuGFETs. We have elucidated how the L-BTBT and T-BTBT GIDL components manifest themselves in the emerging MuGFETs. We also discussed the implications of utilizing a high-*k* dielectric in the gate oxide of emerging MuGFETs from a GIDL perspective. While the presence of high-*k* dielectric in the gate oxide increases the leakage current, it also improves the ON-current simultaneously. Therefore, high-*k* gate dielectric would be useful for high-speed applications where high-performance (HP) flavor of FETs are required. On the other hand, while the use of high-*k* dielectric in the gate side-wall spacer reduces the L-BTBT GIDL for conventional MuGFETs, the presence of high-*k* gate-sidewall spacer in the underlap channel MuGFETs increases the GIDL current. Therefore, the gate-sidewall spacer material can be used as an additional knob for optimizing the performance of emerging MuGFETs. However, the application of high-*k* dielectric degrades the dynamic performance of the emerging MuGFETs. Therefore, there is a trade-off between the improvement in the static performance from a GIDL perspective and the degradation in the speed of the emerging MuGFETs. The design guidelines provided in this chapter may enable the device designers to realize highly energy-efficient MuGFETs for future technology nodes and enable the development of a green society.

REFERENCES

[1] C. Auth et al., "A 22nm high performance and low-power CMOS technology featuring fully-depleted tri-gate transistors, self-aligned contacts and high density MIM capacitors," *2012 Symposium on VLSI Technology (VLSIT)*, Honolulu, HI, pp. 131–132, 2012. doi:10.1109/VLSIT.2012.6242496

[2] J. M. Rabaey, A. P. Chandrakasan, and B. Nikolic, *Digital Integrated Circuits*, Englewood Cliffs, NJ: Prentice Hall, 2002.

[3] A. Chaudhry and M. J. Kumar, "Controlling short-channel effects in deep submicron SOI MOSFETs for improved reliability: A review," *IEEE Trans. Dev. Mater. Rel.*, vol. 4, pp. 99–109, 2004.

[4] S. Sahay and M. J. Kumar. *Junctionless Field-effect Transistors: Design, Modeling, and Simulation.* New York: John Wiley & Sons, 2019.

[5] V. Nathan and N. C. Das, "Gate-induced drain leakage currents in MOS devices," *IEEE Trans. Electron Devices*, vol. 40, no. 10, pp. 1888–1890, 1993.

[6] T. Hoffmann, G. Doornbos, I. Ferain, N. Collaert, P. Zimmerman, M. Goodwin, R. Rooyackers, A. Kottantharayil, Y. Yim, A. Dixit, K. De Meyer, M. Jurczak, and S. Biesemans, "GIDL (gate induced drain leakage) and parasitic Schottky barrier leakage elimination in aggressively scaled HfO2/TiN FinFET devices," *IEDM Tech. Dig.*, pp. 725–729, 2005.

[7] M. J. Kumar and S. Sahay, "Controlling BTBT induced parasitic BJT action in junctionless FETs using a hybrid channel," *IEEE Trans. Electron Devices*, vol. 63, no. 8, pp. 3350–3353, 2016.

[8] S. Sahay and M. J. Kumar, "Realizing efficient volume depletion in SOI junctionless FETs," *IEEE J. Electron Devices Soc.*, vol. 4, no. 3, pp. 110–115, 2016.

[9] S. Sahay and M. J. Kumar, "Symmetric operation in an extended back gate JLFET for scaling to the 65 nm regime considering quantum confinement effects," *IEEE Trans. Electron Devices*, vol. 64, no. 1, pp. 21–27, 2017.

[10] S. Ogura, P. J. Tsang, W. W. Walker, D. L. Critchlow, and J. F. Shepard, "Design and characteristics of the lightly doped drain–source (LDD) insulated gate field-effect transistor," *IEEE Trans. Electron Devices*, vol. 27, no. 8, pp. 1359–1367, 1980.

[11] H. Park and B. Choi. "Gate-induced drain leakage current of MOSFET with junction doping dependence," *2011 International Semiconductor Device Research Symposium (ISDRS)*. IEEE, 2011.

[12] H. Zhu, Semiconductor Nanowire MOSFETs and Applications. In *Nanowires New Insights*, vol. 101, 2017.

[13] D. Nagy et al., "FinFET versus gate-all-around nanowire FET: Performance, scaling, and variability," *IEEE J. Electron Devices Soc.*, vol. 6, pp. 332–340, 2018.

[14] P. Zheng et al., "FinFET evolution toward stacked-nanowire FET for CMOS technology scaling," *IEEE Trans. Electron Devices*, vol. 62, no. 12, pp. 3945–3950, 2015.

[15] J. B. Chang, C.-H. Lin, and J. W. Sleight, "Nanowire fet and finfet," U.S. Patent Application No. 13/548,554.

[16] R. Hajare, C. Lakshminarayana, and G.H. Raghunandan. "Performance enhancement of FINFET and CNTFET at different node technologies," *Microsyst. Technol.*, vol. 22, no. 5, pp. 1121–1126, 2016.

[17] S. Sahay and M. J. Kumar, "Spacer design guidelines for nanowire FETs from gate-induced drain leakage perspective," *IEEE Trans. Electron Devices*, vol. 64, no. 7, pp. 3007–3015, 2017.

[18] T. S. A. Samuel, N. Arumugam, and A. Shenbagavalli, "Drain current characteristics of silicon nanowire field effect transistor," *ICTACT J. Microelectron.*, vol. 2, no. 3, pp. 284–287, 2016.

[19] E. Gnani et al., "Theory of the junctionless nanowire FET," *IEEE Trans. Electron Devices*, vol. 58, no. 9, pp. 2903–2910, 2011.

[20] X. Chen and C.M. Tan, "Modeling and analysis of gate-all-around silicon nanowire FET," *Microelectron. Reliab.*, vol. 54, no. 6–7, pp. 1103–1108, 2014.

[21] J. W. Sleight et al. "Gate-all-around silicon nanowire MOSFETs and circuits," *68th Device Research Conference.* IEEE, 2010.

[22] D. Sacchetto et al., "Fabrication and characterization of vertically stacked gate-allaround Si nanowire FET arrays," *2009 Proceedings of the European Solid State Device Research Conference.* IEEE, 2009.

[23] J. Fan, M. Li, X. Xu, Y. Yang, H. Xuan, and R. Huang, "Insight into gate-induced drain leakage in silicon nanowire transistors," *IEEE Trans. Electron Devices*, vol. 62, no. 1, pp. 213–219, 2015.

[24] S. Sahay and M. J. Kumar, "Physical insights into the nature of gate-induced drain leakage in ultrashort channel nanowire FETs," *IEEE Trans. Electron Devices*, vol. 64, no. 6, pp. 2604–2610, 2017.

[25] S. Sahay and M. J. Kumar, "A novel gate-stack-engineered nanowire FET for scaling to the sub-10-nm regime," *IEEE Trans. Electron Devices*, vol. 63, no. 12, pp. 5055–5059, 2016.

[26] S. Sahay and M. J. Kumar, "Controlling L-BTBT and volume depletion in nanowire JLFETs using core-shell architecture," *IEEE Trans. Electron Devices*, vol. 63, no. 9, pp. 3790–3794, 2016.

[27] S. Sahay and M. J. Kumar, "Nanotube junctionless FET: proposal, design, and investigation," *IEEE Trans. Electron Devices*, vol. 64, no. 4, pp. 1851–1856, 2017.

[28] S. Sahay and M. J. Kumar, "Realizing efficient volume depletion in SOI junctionless FETs," *IEEE J. Electron Devices Soc.*, vol. 4, no. 3, pp. 110–115, 2016.

[29] T. Hoffmann, G. Doornbos, I. Ferain, N. Collaert, P. Zimmerman, M. Goodwin, R. Rooyackers, A. Kottantharayil, Y. Yim, A. Dixit, K. De Meyer, M. Jurczak, and S. Biesemans, "GIDL (gate induced drain leakage) and parasitic Schottky barrier leakage elimination in aggressively scaled HfO$_2$/TiN FinFET devices," *IEDM Tech. Dig.*, pp. 725–729, 2005.

[30] M. J. Kumar and S. Sahay, "Controlling BTBT induced parasitic BJT action in junctionless FETs using a hybrid channel," *IEEE Trans. Electron Devices*, vol. 63, no. 8, pp. 3350–3353, 2016.

[31] S. Sahay and M. J. Kumar, "Insight into lateral band-to-band-tunneling in nanowire junctionless FETs," *IEEE Trans. Electron Devices*, vol. 63, no. 10, pp. 4138–4142, 2016.

[32] J. Hur, B.-H. Lee, M.-H. Kang, D.-C. Ahn, T. Bang, S.-B. Jeon, and Y.-K. Choi, "Comprehensive analysis of gate-induced drain leakage in vertically stacked nanowire FETs: Inversion-mode vs. junctionless mode," *IEEE Electron Device Lett.*, vol. 37, no. 5, pp. 541–544, 2016.

[33] A. B. Sachid et al., "Sub-20 nm gate length FinFET design: Can high-spacers make a difference?" *IEDM Tech. Dig.*, pp. 697–700, 2008.

[34] A. B. Sachid, M.-C. Chen, and C. Hu, "FinFET with high- spacers for improved drive current," *IEEE Electron Device Lett.*, vol. 37, no. 7, pp. 835–838, 2016.

[35] J.-H. Hong et al., "Impact of the spacer dielectric constant on parasitic RC and design guidelines to optimize DC/AC performance in 10-nm-node Si-nanowire FETs," *Jpn. J. Appl. Phys.*, vol. 54, pp. 04DN05-1–04DN05-5, 2015.

[36] J.-S. Yoon, K. Kim, T. Rim, and C.-K. Baek, "Performance and variations induced by single interface trap of nanowire FETs at 7-nm node," *IEEE Trans. Electron Devices*, vol. 64, no. 2, pp. 339–345, 2017.

[37] J.-S. Yoon et al., "Junction design strategy for Si bulk FinFETs for system-on-chip applications down to the 7-nm node," *IEEE Electron Dev. Lett.*, vol. 36, no. 10, pp. 994–996, 2015.

[38] S. Sahay and M. J. Kumar, "Comprehensive analysis of gate-induced drain leakage in emerging FET architectures: Nanotube FETs versus nanowire FETs," *IEEE Access*, vol. 5, pp. 18918–18926, 2017.

[39] D. Tekleab, "Device performance of silicon nanotube field-effect transistor," *IEEE Electron Device Lett.*, vol. 35, no. 5, pp. 506–508, 2014.

[40] A. N. Hanna, H. M. Fahad, and M. M. Hussain, "InAs/Si hetero-junction nanotube tunnel transistors," *Sci. Rep.*, vol. 9, p. 9843, 2015.

[41] H. M. Fahad and M. M. Hussain, "High-performance silicon nanotube tunneling FET for ultralow-power logic applications," *IEEE Trans. Electron Devices*, vol. 60, no. 3, pp. 1034–1039, 2013.

[42] A. N. Hanna and M. M. Hussain, "Si/Ge hetero-structure nanotube tunnel field-effect transistor," *J. Appl. Phys.*, vol. 117, no. 1, p. 014310, 2015.

[43] A.K. Jain, S. Sahay, and M.J. Kumar, "Controlling L-BTBT in emerging nanotube FETs using dual-material gate," *IEEE J. Electron Dev. Soc.*, vol. 6, pp. 611–621, 2018.

[44] S. Tayal and Nandi. "Study of 6T SRAM cell using high-k gate dielectric based junctionless silicon nanotube FET," *Superlattice. Microst.*, vol. 112, pp. 143–150, 2017.

[45] H. M. Fahad, C. E. Smith, J. P. Rojas, and M. M. Hussain, "Silicon nanotube field-effect transistor with core–shell gate stacks for enhanced high-performance operation and area scaling benefits," *Nano Lett.*, vol. 11, no. 10, pp. 4393–4399, 2011.

[46] H. M. Fahad and M. M. Hussain, "Are nanotube architectures advantageous than nanowire architectures for field-effect transistor applications?" *Sci. Rep.*, vol. 2, no. 2, p. 475, 2012.

[47] D. Tekleab, H. H. Tran, J. W. Sleight, and D. Chidambarrao, "Silicon nanotube MOSFET," U.S. Patent 0 217 468, August 30, 2012.

[48] S. Tayal and A. Nandi, "Optimization of gate-stack in junctionless Si nanotube FET for analog/RF applications," *Mater. Sci. Semicon. Proc.*, vol. 80, pp. 63–67, 2018.

[49] L. Ding, Z. Wang, T. Pei, Z. Zhang, S. Wang, H. Xu, F. Peng, Y. Li, and L.-M. Peng, "Self-aligned U-gate carbon nanotube field-effect transistor with extremely small parasitic capacitance and drain-induced barrier lowering," *ACS Nano*, vol. 5, no. 4, pp. 2512–2519, 2011.

[50] S. Tayal and A. Nandi, "Analog/RF performance analysis of inner gate engineered junctionless Si nanotube," *Superlattice. Microst.*, vol. 111, pp. 862–871, 2017.

[51] S. Tayal and A. Nandi, "Interfacial layer dependence of high-k gate stack based conventional trigate FinFET concerning analog/RF performance," In *2018 4th International Conference on Devices, Circuits and Systems (ICDCS)*, pp. 305–308. IEEE, 2018.

[52] S. Tayal and A. Nandi, "Analog/RF performance analysis of channel engineered high-k gate-stack based junctionless trigate-FinFET," *Superlattice. Microst.*, vol. 112, pp. 287–295, 2017.

[53] A.B. Sachid and C. Hu, "A little-known benefit of FinFET over planar MOSFET in highperformance circuits at advanced technology nodes," *2012 IEEE International SOI Conference (SOI)*. IEEE, 2012.

[54] H. Mertens et al., "Gate-all-around MOSFETs based on vertically stacked horizontal Si nanowires in a replacement metal gate process on bulk Si substrates," *2016 IEEE Symposium on VLSI Technology*. IEEE, 2016.

[55] R.S. Pal, S. Sharma, and S. Dasgupta, "Recent trend of FinFET devices and its challenges: A review," *2017 Conference on Emerging Devices and Smart Systems (ICEDSS)*. IEEE, 2017.

[56] K.J. Kuhn, "Considerations for ultimate CMOS scaling," *IEEE Trans. Electron Devices*, vol. 59, no. 7, pp. 1813–1828, 2012.

[57] D. Bhattacharya and N.K. Jha, "FinFETs: From devices to architectures," *Adv. Electron.*, vol. 2014, 2014.

[58] D.M. Kim and Y.-H. Jeong, eds. *Nanowire Field Effect Transistors: Principles and Applications*. Vol. 43. Berlin, New York: Springer, 2014.

[59] S. Barraud et al., "Performance and design considerations for gate-all-around stacked-nanowires FETs," *2017 IEEE International Electron Devices Meeting (IEDM)*. IEEE, 2017.

[60] A. Veloso et al., "Nanowire nanosheet FETs for ultra-scaled, high-density logic and memory applications," *Solid-State Electron.*, vol. 168, p. 107736, 2020.

[61] D. Jang et al., "Device exploration of nanosheet transistors for sub-7-nm technology node," *IEEE Trans. Electron Devices*, vol. 64, no. 6, pp. 2707–2713, 2017.

[62] N. Loubet et al., "Stacked nanosheet gate-all-around transistor to enable scaling beyond FinFET," *2017 Symposium on VLSI Technology*, pp. T230–T231, 2017.

[63] P. Ye, T. Ernst, and M. V. Khare, "The last silicon transistor: Nanosheet devices could be the final evolutionary step for Moore's Law," *IEEE Spectr.*, vol. 56, no. 8, pp. 30–35, 2019, doi:10.1109/MSPEC.2019.8784120.

[64] M.M.R. Moayed, T. Bielewicz, H. Noei, A. Stierle, and C. Klinke, "High-performance n and p-type field-effect transistors based on hybridly surface-passivated colloidal PbS nanosheets," *Adv. Funct. Mater.*, vol. 28, no. 19, p. 1706815, 2018.

[65] D. Ryu et al., "Investigation of gate sidewall spacer optimization from OFF-state leakage current perspective in 3-nm node device," *IEEE Trans. Electron Devices*, vol. 66, no. 6, pp. 2532–2537, 2019.

[66] H. Kim et al., "Strain engineering for 3.5-nm node in stacked-nanoplate FET," *IEEE Trans. Electron Devices*, vol. 66, no. 7, pp. 2898–2903, 2019. doi:10.1109/TED.2019.2917503.

[67] H. Kim et al., "Optimization of stacked nanoplate FET for 3-nm node," *IEEE Trans. Electron Devices*, vol. 67, no. 4, pp. 1537–1541, 2020. doi:10.1109/TED.2020.2976041.

[68] Y. Choi et al., "Simulation of the effect of parasitic channel height on characteristics of stacked gate-all-around nanosheet FET," *Solid-State Electron.*, vol. 164, p. 107686, 2020.

[69] D. Ryu, M. Kim, J. Yu, S. Kim, J. Lee, and B. Park, "Investigation of sidewall high-k interfacial layer effect in gate-all-around structure," *IEEE Trans. Electron Devices*, vol. 67, no. 4, pp. 1859–1863, 2020. doi:10.1109/TED.2020.2975255.

[70] M.-W. Ma et al., "Impact of high-k offset spacer in 65-nm node SOI devices," *IEEE Electron Device Lett.*, vol. 28, no. 3, pp. 238–241, 2007. doi:10.1109/LED.2007.891282.

[71] A.B. Sachid, M.-C. Chen, and C. Hu, "FinFET with high-k spacers for improved drive current," *IEEE Electron Device Lett.*, vol. 37, no. 7, pp. 835–838, 2016. doi:10.1109/LED.2016.2572664.

[72] C. Yin, P. C. H. Chan, and M. Chan, "An air spacer technology for improving short-channel immunity of MOSFETs with raised source/drain and high-gate dielectric," *IEEE Electron Device Lett.*, vol. 26, no. 5, pp. 323–325, 2005. doi:10.1109/LED.2005.846584.

[73] A.B. Sachid et al., "FinFET with encased air-gap spacers for highperformance and lowenergy circuits," *IEEE Electron Device Lett.*, vol. 38, no. 1, pp. 16–19, 2017. doi:10.1109/LED.2016.2628768.

[74] H. Ko et al., "Device investigation of nanoplate transistor with spacer materials," *IEEE Trans. Electron Devices*, vol. 66, no. 1, pp. 766–770, 2018.

[75] D. Ryu et al., "Design and optimization of triple-k spacer structure in two-stack nanosheet FET from OFF-state leakage perspective," *IEEE Trans. Electron Devices*, vol. 67, no. 3, pp. 1317–1322, 2020. doi:10.1109/TED.2020.2969445.

6 Advanced FET Design Using High-*k* Gate Dielectric and Characterization for Low-Power VLSI

P. Vimala
Dayananda Sagar College of Engineering, Bangalore, India

T. S. Arun Samuel
National Engineering College, Kovilpatti, India

CONTENTS

6.1 INTRODUCTION

The integrated circuit (IC) is a semiconductor wafer, often referred to as a microchip, on which millions of small transistors, capacitors, and resistors are fabricated. Due to its high current carrying ability compared to bipolar transistors, the field effect transistor (FET) plays a significant role in digital circuits. The field-effect transistor (FET) is a transistor that regulates charge carriers' movement by using an electric field. As they require single-type charge carriers for FET operation, it is also recognized as unipolar transistors. The FET transistor types are varying entirely from the basics of BJT transistors. Gate, drain and source are the three terminals used in FET devices. They are used in RF electronic circuits for switching and power amplification. In general, field effect transistors are classified as junction FETs (JFETs) and metal oxide semiconductor FETs (MOSFETs).

6.2   JFETS AND MOSFETS

A type of field-effect transistor used as an electrically operated switch is the junction FET transistor (JFET). Electrical energy flows between sources to drain terminals via an active channel. The channel is stressed by applying a reverse bias voltage to the gate terminal to entirely turn off the electric current. JFET can be divided based on its activity into two groups: N-type and JFET P-type. JFET output decreases as frequency increases due to internal capacitance feedback. JFETs can be used only in the depletion mode.

MOSFET stands for the thin insulating layer of the silicon oxide layer in the area between the metal and the semiconductor. The name "field-effect transistor," indicates the characteristics of this device. MOSFETs also have two types of enhancement mode: MOSFET (E-MOSFET) and depletion mode MOSFET (D-MOSFET). These two groups are also categorized into the forms of n-channel and p-channel.

Except for the symbols, the structure of the JFET varies from the MOSFET structure. Examples of n-channel JFET and n-channel MOSFET shown in Figure 6.1 have taken for discussion.

In Figure 6.1, the JFET body is made up of n-type semiconductor, and the MOSFET body is made up of p-type semiconductor, also recognized as a p-type

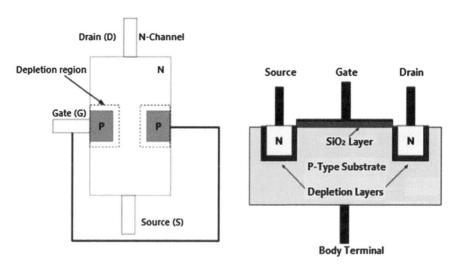

FIGURE 6.1   Structure of JFET and MOSFET.

substrate. JFET has three terminals: the gate, the drain, and the source. The MOSFET also has the same three terminals additionally with another terminal named as the body. This terminal is not essential in MOSFET switching activities, as it is internally linked to the source.

JFET has the doping of p-type materials on opposite sides, and both p-type doping regions are connected via the gate terminal. The gate terminal is not separated from the JFET's body. In the case of MOSFET, the gate is not explicitly attached to the body. The SiO_2 dielectric is used for isolating the gate terminal to the MOSFET body.

MOSFET is the most commonly used device and has a profound influence on IC technology.

6.3   SCALING AND SHORT CHANNEL EFFECTS (SCES)

Among all currently available IC technologies, CMOS is found to have the most significant advantages. In CMOS technology, continuous advances can be made, and these advances are managed and sustained by constant CMOS scaling, commonly referred to as the Moore's law. The concept of scaling theory is described in Figure 6.2.

The method of decreasing the field-effect transistors' size and interconnectedness is called CMOS scaling without compromising the IC's functionality. The gate loses its complete power over its potential when the channel length of a traditional MOSFET decreases. The different short channel effects (SCEs) and degradation mechanisms causing transistor efficiency and reliability become more closely related as the device dimensions are scaled down. The temporary channel effects result in various problems, including drain-induced barrier lowering (DIBL), saturation in velocity, threshold voltage roll-off, rise in reverse leakage current, hot carrier effects, mobility reduction, etc. The device output is reduced and increases the leakage current due to these SCEs [1, 2]. A variety of alternative device architectures have been proposed to solve the impact of the SCEs.

With the continuous efforts by researchers to improve the device outcome and better control of SCEs, MOS transistors have migrated from single-gate devices to three-dimensional devices with multiple-gate structures.

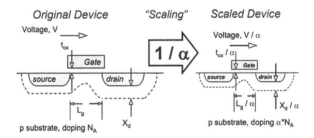

FIGURE 6.2   Scaling theory caption.

6.4   MULTI-GATE MOSFETS

The multi-gate structures broadly classified as three types: double-gate (DG), tri-gate (TG), and quad-gate, are shown in Figure 6.3. The double-gate structure refers to the gate material present on two opposite sides of the device [3]. One significant discussion about this double-gate structure is that the two gate terminals can bias together or separately. Based on this biasing difference, the double-gate structures are named symmetric and asymmetric structures, shown in the Figure 6.3a and b. Similar to double gate, a tri-gate structure [4] is designed with the gate material folded on three sides of the device, and for surrounding gate [5], the gate material folded all sides of the devices, also known as a gate-all-around (GAA) structure.

6.5   LEAKAGE CURRENT

Silicon dioxide (SiO_2) has been used as a dielectric material for gate oxides for decades. The thickness of the dielectric material has gradually decreased to increase the gate's capacitance and drive current, improving the efficiency of the device. When the thickness of dielectric material scales reaches less than 2 nm, leakage currents due to tunneling increase significantly, leading to high power consumption and decreased device performance. The scaling of the device threshold voltage (Vth) is necessary to maintain a proper gate over the drive. The decrease in Vth leads to an overall rise in the current of the sub-threshold region. Oxide thickness must also be scaled down to control the SCEs and to preserve the transistor drive power at a low input voltage. High tunneling current results from vigorous oxide thickness scaling via the transistor gate [6]. Scaled devices often involve the use of a higher density of substrate doping. Under high reversed biassing, it induces significant leakage current through all the device junctions. Figure 6.4 shows the three main leakage mechanisms. The following are the three primary forms of pathways for leakage: gate oxide leakage, subthreshold leakage, and BTBT (band-to-band tunneling) leakage, where the BTBT is also known as reverse-bias PN-junction leakage

6.6   IMPORTANCE OF HIGH-*k* MATERIALS

The key to MOSFET's success is also even said: SiO_2/Si interface properties are excellent. However, to suppress the excess leakage current, high-*k* materials must

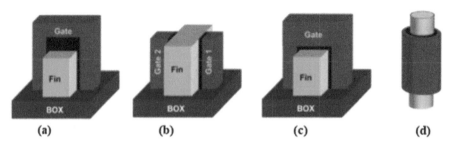

FIGURE 6.3   Multi-gate structures: (a) symmetric DGMOS; (b) asymmetric DGMOS; (c) TGMOS; (d) GAAMOS.

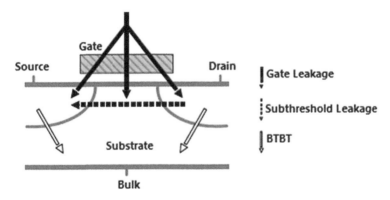

FIGURE 6.4 Three main leakage mechanisms of MOSFET.

be added as gate dielectrics. Adding a high-*k* material with the silicon dioxide enables improved capacitance of the gate without any leakage current. The equivalent oxide thickness (EOT) of high-*k* dielectric materials is 1.0 nm with minimal gate oxide leakage, the acceptable transistor threshold voltage for MOSFETs, and transistor channel mobility similar to SiO_2. Feasible high-*k* materials are Si_3N_1 (dielectric value of 7.5), Al_2O_3 (dielectric value of 10), $LaAlO_3$ (dielectric value of 15), HfO_2/ZrO_2 (dielectric value of 25), La_2O_3 (dielectric value of 27), and TiO_2 (dielectric value as 40).

Table 6.1 summarizes the list of high-*k* materials.

TiO_2 has a high dielectric value of 40. But it has a high OFF current because of its less energy bandgap compare to other high-*k* materials. One of the best alternative high-*k* materials is ZrO_2. It has less physical thickness because of its nano-size substance and decreases the direct tunneling of current.

The high-*k* materials can be introduced in two different ways of configurations in the MOSFET structure [7]. The first one is a high-*k* stack with SiO_2 and half gate length high-*k* dielectric configuration shown in Figure 6.5. A thicker dielectric over the substrate is needed for the stacked dielectric. Another gate-engineered structure is implemented to decrease density and increase efficiency. Here, the first half of the gate length is with SiO_2 dielectric, and the second half is with HfO_2.

TABLE 6.1
List of High-*k* Materials

S. No	High-*k* Materials	High-*k* Values	Energy Gap
1	SiO_2	3.9	9
2	Si_3N_1	7.5	5.3
3	Al_2O_3	10	6
4	$LaAlO_3$	15	5.6
5	ZrO_2	25	5.8
6	HfO_2	25	6
7	TiO_2	40	3.5

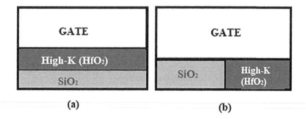

FIGURE 6.5   High-*k* material: (a) stacked configuration; (b) half gate length configuration.

6.7   NEXT-GENERATION TRANSISTORS

Every new generation of integrated circuits (ICs) is expected to have increased flex-ibility, decreased cost per function, and improved performance. The most important way to achieve these merits is by downscaling MOS transistors. Although the IC industries prefer to continue the scaling of advanced MOS transistors as much as possible, the ultra-thin gate oxide of the transistor produces unacceptable gate leak-age current. Due to the possible use of different dielectrics in MOS transistors, oxide materials with high permittivity dielectric materials (high-*k*) in the gate stack have attracted much interest in order to suppress gate leakage current. Semiconductor industries are looking for new alternate devices due to the end result of scaling of the MOS transistor's physical limit. Tunnel field effect transistors (TFETs), junctionless transistors, graphene field effect transistors (G-FETs), and carbon nanotubes are the new next -eneration devices under research. The impacts of high-*k* material used in next-generation transistors are as follows.

6.8   TUNNEL FILED EFFECT TRANSISTORS WITH HIGH-*k* DIELECTRIC

TFETs are the most promising future nanodevices for low power application. The fundamental restriction of CMOS devices is its sub-threshold swing (SS), which is limited to 60mV/dec. However, the SS of the TFET devices could be reduced below 60mV/dec, and its OFF-current is approximately picoampere to femtoampere. In spite of these advantages in TFET, there are certain drawbacks present in the device, specifi-cally its low ON-current, ambipolar behavior, and gate leakage current. To overcome these drawbacks, TFETs have been designed with various techniques like multi-gate TFETs [8], heterojunction TFETs [9, 10], and hetero-dielectric TFETs [11–13].

Gate leakage current has become one of the key parameter expected to affect the sub-threshold characteristics of TFETs. In general, the oxide capacitance is expressed as $C_{ox} = \varepsilon_{ox}/t_{ox}$. Here the ε_{ox} permittivity of the oxide layer and t_{ox} is the thickness of the oxide layer. From this capacitance expression, it is clear that a higher oxide capacitance results in a greater charge generated in the channel region, thereby increasing the gate control over the electrostatics of the channel. A smaller gate oxide thickness may nevertheless lead to carriers from the channel tunneling to the gate terminal, generating an unwanted gate leakage current. However, this can be elimi-nated by adding a high-permittivity materials (high-*k*) in addition to the conventional thin silicon dioxide layer; it can also increase gate control over the channel region.

By varying the structural parameters of the TFET, including high-*k* dielectric, results in a significant improvement in device characteristics. Let us review the certain vital high-*k* dielectric tunnel FET structures.

6.9   DG TFET WITH HIGH-*k* DIELECTRIC

Several research works have been carried out to resolve the gate leakage current and loss of gate control over the channel in TFETs. When the device is scaled below nanometer regime, the gate losses its control over the channel region (intrinsic region), and it is not possible to control the switching characteristics. However, the TFET device acts as a normal p-i-n diode. Hence, to maintain gate control over the channel for a short channel device, high-*k* dielectric material is placed instead of conventional SiO$_2$ dielectric. One such device is DG TFET with high-*k* dielectric, proposed by Boucart et al. [14]. The schematic diagram of high-*k* dielectric material used in double-gate (DG) TFET is shown in Figure 6.6. The basic structure is a p-i-n diode with front and back gates and the dielectric material used are Si$_3$N$_4$, HfO$_2$, and ZrO$_2$. As the gate bias is increased, the tunneling takes place between the P+ source and intrinsic channel region. The outcome of this study was the increase in the ON-current when the gate dielectric constant increased. The DG TFET with high-*k* dielectric provides better ON-current of 0.23 mA/V$_{GS}$ = 1.8V and gives a slight improvement in the average sub-threshold swing at 57 mV/dec. The tunneling probability has been analytically expressed as [15],

$$T(E) \propto \exp\left(\frac{4\sqrt{2m*}E_g^{3/2}}{3|e|\hbar(E_g+\Delta\Phi)} \sqrt{\frac{\varepsilon_{Si}}{\varepsilon_{ox}}} t_{ox}t_{Si} \right) \Delta\Phi \tag{6.1}$$

It is clearly inferred from Equation (6.1) that as the high-*k* dielectric constant value increases, the tunneling of charge carriers increases, which leads to improved

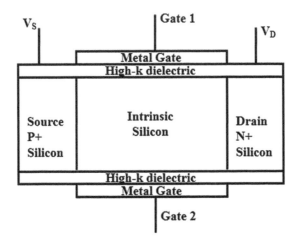

FIGURE 6.6   Schematic diagram of DG TFET with high-*k* dielectric.

ON-current. Even though high-*k* is having more merits when the high-*k* materials are directly put over a silicon channel, they will not have good bonding with each other. Therefore, stacking the dielectric SiO_2 and high-*k* dielectric material in this fashion can greatly enhance the SS characteristics.

6.10   TRI-GATE TFET WITH HIGH-*k* DIELECTRIC

Different high-*k* materials used in Tri-Gate TFET has been studied by Vimala et al. [12]. Many investigators were keen to build models for single-gate tunnel FET to analyze device physics for further development and scaling. Fortunately, single-gate TFETs have low ON-state current compared to traditional MOSFETs. This encourages investigators to build tri-gate TFET model that provide enhanced scaling capacity, reduced SCEs, and improved SS. To avoid the gate leakage current and improve the ON-current of the device, different high-*k* materials are adopted to study and simulate the performance of the tri-gate TFET. The I_D-V_{GS} characteristics of different high-*k* materials used in tri-gate TFET are shown in Figure 6.7.

It is clearly seen from the Figure 6.7 that TiO_2 used as high-*k* dielectric displays a greater ON-current compared to other high-*k* dielectric materials. Comparison of the tri-gate TFET with single-material (SM) tri-gate TFET and dual-material (DM) tri-gate TFET is also provided. Note that the tri-gate TFET device has an output drain

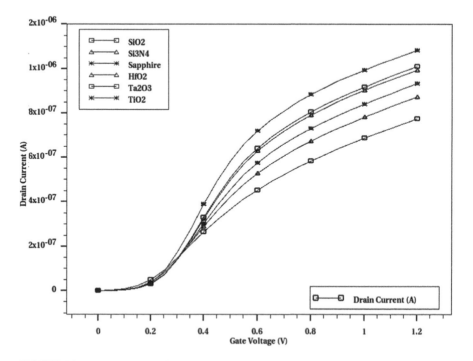

FIGURE 6.7 I_D-V_{GS} characteristics of tri-gate TFET with different dielectric materials with different high-*k* dielectric.

current of 7.25 µA and that the output has increased by 45% compared to the SM tri-gate TFET device with only 5 µA output.

6.11   DG TFET WITH SIO$_2$/HIGH-k STACKED DIELECTRIC

To avoid the interface defects (lattice mismatching) at the silicon channel and high-k interface, Kumar et al. [16] proposed a SiO$_2$/high-k stacked DG TFET device structure. The schematic diagram of SiO$_2$/high-k stacked DG TFET is shown in Figure 6.8. As the thickness of the SiO$_2$/high-k stacked DG TFET increases, the threshold voltage (V_{TH}) decreases, which leads to a decrease in energy band gap between the source/channel interfaces.

The simulation results reveal that the ON-current of SiO$_2$/high-k stacked DG TFET structure is 10^{-6}A/µm and the OFF-current is 10^{-16}A/µm, while the SiO$_2$ dielectric DG TFET structure provides the ON-current of 10^{-8}A/µm and the OFF-current of 10^{-16}A/µm. Hence, the impact of gate leakage current is greatly reduced in the high-k stacked DG TFET structure.

6.12   HETEROJUNCTION TFET WITH SIO$_2$/HIGH-k STACKED DIELECTRIC

The impact of SiO$_2$/high-k stacked dielectric in heterojunction TFET has been analyzed by Vimala et al. [10]. Hafnium oxide (HfO$_2$) is used as the high-k material, and three different work function materials are used at the gate. The schematic diagram of heterojunction TFET is shown in Figure 6.9 where the source is designed using germanium and the channel/drain is designed using silicon. The efficient tunneling takes place at the source/channel heterojunction.

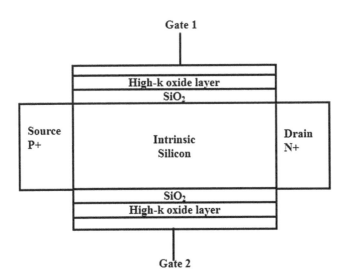

FIGURE 6.8   Schematic diagram of DG TFET with SiO$_2$/high-k dielectric.

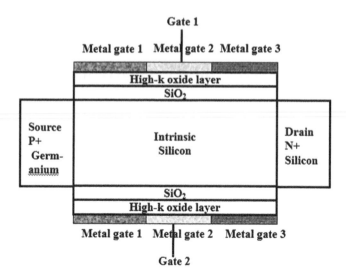

FIGURE 6.9   Schematic diagram of heterojunction TFET with SiO₂/high-*k* dielectric.

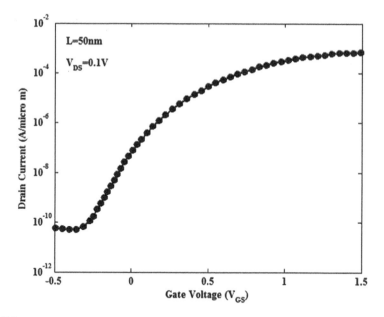

FIGURE 6.10 I_D-V_{GS} characteristics of heterojunction SiO₂/high-*k* stacked TFET.

The simulation results of high-*k* dielectric placed over the SiO₂ dielectric show better gate control over the channel region (intrinsic region) in terms of surface potential and electric field. Peak electric field is present at the tunneling junction, which helps improve the ON-current. Figure 6.10 shows the I_D-V_{GS} characteristics of heterojunction SiO₂/high-*k* stacked TFET. Figure shows that, due to the stacked dielectric heterojunction structure, the ON-current has been improved to 10^{-4}A/μm and the OFF-current to approximately 10^{-11}A/μm.

6.13   JUNCTIONLESS TRANSISTORS WITH HIGH-*k* DIELECTRIC

In MOSFET technology, the p-n junction is the key element that separates the substrate from the device and constrains the flow of current in the inversion region. In short-channel devices, the development of source–channel and drain–channel interfaces requires an ultra-steep doping profile, which severely complicates manufacturing. In nanometer scale, control over source/drain junction depth is crucial. In order to get rid of these limitations, a new device with technologically and economically benefits for the semiconductor industry is a junctionless transistor (JL).

The schematic view of a junctionless transistor is shown in Figure 6.11. It consists of a gate to regulate the carrier (channel) concentration by applying a field (electric) in a heavily doped silicon film; hence the resistivity of the film is altered. Silicon dioxide (SiO_2) acts as an insulating layer between the gate and the substrate. Generally, uniform heavily doped junctionless devices (i.e., source/drain and channel regions have same type of doping profile), having no metallurgical junction, become very compact to CMOS process accessibility.

The major challenge for a junctionless transistor is to achieve improvement in the ON-current with reduced operating voltage and diminishing the leakage current. This leads to further exploration in terms of novel multi-gate junctionless transistors with high-*k* dielectric. The impact of a high-*k* material on a junctionless double-gate device structure was explored by Baidya et al. [17]. Figure 6.12 depicts the schematic

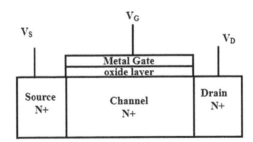

FIGURE 6.11   Schematic diagram of junctionless transistor.

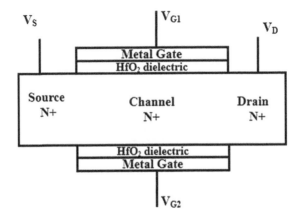

FIGURE 6.12   Schematic high-*k* dielectric double-gate junctionless transistor.

view of a double-gate junctionless transistor with dielectric (high-*k*) as insulator. The drain current value of a double-gate junctionless transistor with high-*k* dielectric applied directly onto the surface of the silicon with hafnium dioxide provides 9 μA for different gate voltages with fixed value of Vds = 0.05 V. Similarly, for silicide nitride and silicon dioxide dielectrics, the values of drain current were found to be 7.5 and 6.7 μA, respectively.

6.14 DOUBLE-GATE JUNCTIONLESS TRANSISTOR SIO₂/HIGH-*k* STACKED DIELECTRIC

The performance of gate stack architecture in a junctionless double gate field effect transistor was discussed by Fabiha et al. [18]. In gate stack architecture, the dielectric material was not placed directly on the surface of the silicon body; instead, it was sandwiched above the conventional silicon dioxide. The variation in the device structure was essential in order to remove the process compatibility issues with the direct intervention of high-*k* on the silicon body. Also the high-*k* dielectric placed directly on the silicon surface creates a fringing effect, which enters the channel region and makes changes in the channel charges. It provides better on-off ratio with the value of 10^3A when channel length was reduced to 30 nm or less. The combination of high-*k* and silicon dioxide mitigates the SCEs such as sub-threshold swing and DIBL. Figure 6.13 shows the detailed view of double-gate high-permittivity dielectric constant as gate stack.

6.15 NANOWIRE FETS WITH HIGH-*k*

Nanowire FETs are possible alternatives for MOSFETs because of their specific electronic structure and reduced carrier dispersion induced by one-dimensional quantum confinement effects [19,20]. The nanowire has a hair-like design, which adds operational flexibility. In nanowire FET, the chemical, thermal, optical,

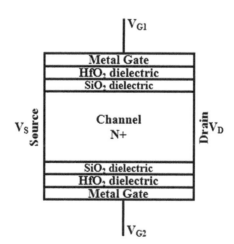

FIGURE 6.13   Schematic high-*k*/SiO₂ dielectric double-gate junctionless transistor.

mechanical, and electronic properties outshine, although these are missing in bulk materials. The structure of a silicon nanowire FET with high-*k* is shown in Figure 6.14.

Figure 6.15 indicates the drain current variation with gate voltage for both SiO_2 and high-*k* compared. HfO_2 is considered as high-*k* material in the device structure. The variations were analyzed for two different Vd values as 0.05 V and 1 V. As per the simulation results, nanowire FET with HfO_2 performs better than SiO_2 with a higher charge density and higher electrostatic field value of HfO_2.

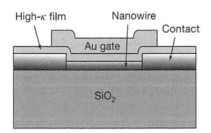

FIGURE 6.14   Structure of nanowire FET with high-*k* material.

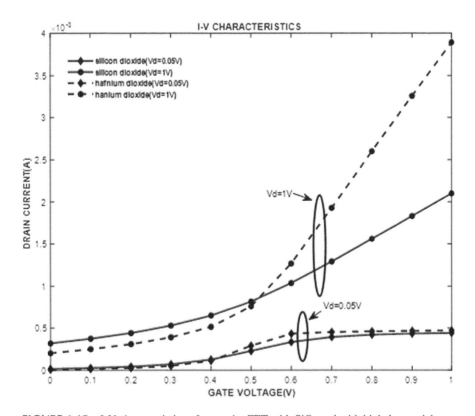

FIGURE 6.15   I-V characteristics of nanowire FET with SiO_2 and with high-*k* material.

6.16   CARBON NANOTUBES FETS WITH HIGH-*k*

Among the various qualities of CNTs and graphene, a few significant reasons to use CNTs and graphene are their excellent performance, such as minimal short channel effects, increased mobility, and adjustable drive currents. The results of an intensive study effort on the feasibility of using carbon nanotube field effect transistors (CNTFETs) as a replacement of a potential semiconductor are shown in Figure 6.16. In addition to what's displayed, due to their high mobility and low defects, these devices are dedicated. The carbon nanotubes are good candidates for the realization of field-effect transistors (FETs) because of the possibility of building a winding gate that offers better electrostatic channel control, with the right current ratios of ON state to OFF state needed for applications in parallel [21]. Figure 6.17 shows the plot of I-V characteristics for different values of coaxial gate voltage (V_G) by plotting drain current (I_D) in amperes along the *y*-axis and drain voltage (V_D) in volts along *x*-axis. The voltage is varied in steps of 0.05 V. Nanotube length is kept constant at 5 nm. From Figure 6.17 it is observed that the drain current increases as the gate voltage increased from 0.2 to 1.0 V. For V_G equal to 0.2 V, the value of I_D starts saturating from 0.05 V (drain voltage) until 0.4 V and reaches a maximum of 0.4 μA.

6.17   SUMMARY

A brief description of FET, its types, and device-level operation has been presented in this chapter. Issues caused by scaling and short-channel effects result in the evolution of multi-gate devices from one-dimensional planar semiconductor devices—the types of leakage currents produced due to SCEs discussed. High-*k* materials' importance in overcoming the leakage currents is addressed in detail. Advanced device architectures such as tunnel FETs, junctionless FETs, carbon nanotubes FETs, and nanowire FETs have excellent properties to be used in low power applications. Various modifications in the structure of these devices and their applications are also presented.

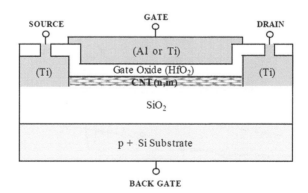

FIGURE 6.16   Structure of CNTFET with high-*k* material.

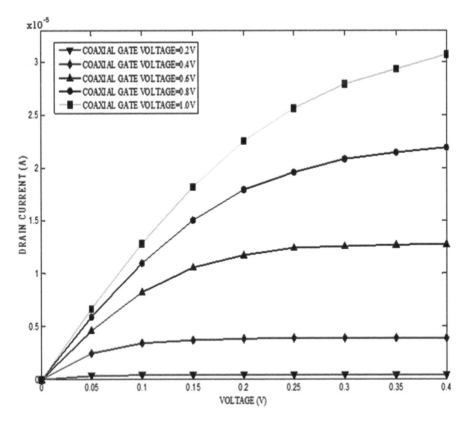

FIGURE 6.17 I_d vs V_d plot for variation over coaxial gate voltage.

REFERENCES

[1] Y. Taur, D. A. Buchanan, W. Chen, D. J. Frank, K. E. Ismail, S.-H. Lo, G. A. Sai-Halasz, R. G. Vishwanathan, H.-J. C. Wann, S. J. Wind, and H.-S. Wong, "CMOS scaling in nanometer regime", *Proceedings of IEEE*, vol. 85, no. 4, pp. 486–504, 1997.

[2] H. Krautschneider, A. Kohlhase, and H. Terletzki, "Scaling down and reliability problems of gigabit CMOS circuits", *Microelectronics Reliability*, vol. 37, no. 1, pp. 19–37, 1997.

[3] P. Vimala and N. B. Balamurugan, "Quantum mechanical compact modeling of symmetric double-gate MOSFETs using variational approach", *Journal of Semiconductors*, vol. 33, no. 3, pp. 034001–034005, 2012.

[4] P. Vimala, "Charge based quantization model for triple-gate FINFETS", *Journal of Nano- and Electronic Physics*, vol. 10, no. 5, pp. 05015, 2018.

[5] P. Vimala and N. R. Nithin Kumar, "Analytical quantum model for germanium channel gate-all-around (GAA) MOSFET", *Journal of Nano Research*, vol. 59, pp. 137–148, 2019.

[6] K. Roy et al. "Leakage current mechanisms and leakage reduction techniques in deep submicrometer CMOS circuits", *Proceedings of IEEE*, vol. 91, no. 2, pp. 305–327, 2003.

[7] V. N. Shrey and N. K. Reddy, "Performance enhancement of multi-gate MOSFETs using gate dielectric engineering", *2018 International Conference on Computing, Power and Communication Technologies (GUCON)*, Greater Noida, India, 2018, pp. 924–928.

[8] S. Komalavalli, T. S. A. Samuel, and P. Vimala, "Performance analysis of triple material tri gate TFET using 3D analytical modelling and TCAD simulation", *AEU – International Journal of Electronics and Communications*, vol. 110, pp. 1–7, 2019.

[9] C. Usha, P. Vimala, T. S. A. Samuel, and M. K. Pandian, "A novel 2-D analytical model for the electrical characteristics of a gate-all-around heterojunction tunnel field-effect transistor including depletion regions", *Journal of Computational Electronics*, online April 2020.

[10] P. Vimala, T. S. A. Samuel, D. Nirmal, and A. K. Panda, "Performance enhancement of triple material double gate TFET with heterojunction and heterodielectric", *Solid State Electronics Letters*, vol. 1, pp. 64–72, 2019.

[11] P. Vimala, K. Netravathi, and T. S. A. Samuel, "Improved drain current characteristics of tunnel field effect transistor with heterodielectric stacked structure", *International Journal of Nano Dimension*, vol. 10, no. 4, pp. 413–419, 2019.

[12] P. Vimala, T. S. A. Samuel, and M. K. Pandian, "Performance investigation of gate engineered tri-gate SOI TFETs with different high-K dielectric materials for low power applications", *Silicon*, vol. 12, pp. 1819–1829, 2020.

[13] I. V. Anand, T. S. A. Samuel, P. Vimala, and A. Shenbagavalli, "Modelling and simulation of hetero-dielectric surrounding gate TFET", *Journal of Nano Research*, vol. 62, pp. 47–58, 2020.

[14] K. Boucart and A. M. Ionescu, "Double-gate tunnel FET with high-k gate dielectric," *IEEE Transactions on Electron Devices*, vol. 54, no. 7, pp. 1725–1733, 2007.

[15] J. Knoch and J. Appenzeller, "A novel concept for field-effect transistors—The tunneling carbon nanotube FET", in *Proceedings of the 63rd DRC*, June 20–22, 2005, vol. 1, pp. 153–156.

[16] S. Kumar, E. Goel, K. Singh, B. Singh, M. Kumar, and S. Jit, "2-D analytical modeling of the electrical characteristics of dual-material double-gate TFETs with a SiO_2/HfO_2 stacked gate-oxide structure", *IEEE Transactions on Electron Devices*, vol. 64, no. 3, pp. 960–968, 2017.

[17] A. Baidyaa, S. Baishyab, and T. R. Lenka, "Impact of thin high-k dielectrics and gate metals on RF characteristics of 3D double gate junctionless transistor", *Materials Science in Semiconductor Processing*, vol. 71, pp. 413–420, 2017.

[18] R. Fabiha, C. N. Saha, & M. S. Islam, "Analytical modeling and performance analysis for symmetric double gate stack-oxide junctionless field effect transistor in subthreshold region", *2017 IEEE Region 10 Humanitarian Technology Conference (R10-HTC)*, Dhaka, pp. 310–313, 2017.

[19] P. L. McEuen, M. S. Fuhrer, and H. K. Park, "Single-walled carbon nanotube electronics", *IEEE Transactions on Nanotechnology*, vol. 1, pp. 78–85, 2002.

[20] P. L. McEuen, M. Bockrath, D. H. Cobden, Y. G. Yoon, and S. G. Louie, "Disorder, pseudospins, and backscattering in carbon nanotubes", *Physical Review Letters*, vol. 83, pp. 5098–5101, 1999.

[21] P. Vimala, S. S. Sharma, M. Bassapuri, and T. Manikanta, "Characteristic analysis of silicon nanowire tunnel field effect transistor (NW-TFET)", *2020 IEEE International Conference on Electronics, Computing and Communication Technologies (CONECCT)*, Bangalore, India, pp. 1–4, 2020, doi:10.1109/CONECCT50063.2020.9198578

7 Simulation and Analysis of Gate Stack DG MOSFET with Application of High-*k* Dielectric Using Visual TCAD

Nisha Yadav, and Sunil Jadav
J.C. Bose UST, YMCA, Faridabad, India

Gaurav Saini
NIT, Kurukshetra, India

CONTENTS

7.1   INTRODUCTION

In 1960, the microelectronics industry had evolved with the invention of MOSFET. Almost every chip began to be fabricated with MOSFET, and the system became popular in the VLSI industry. Gordon Moore in 1965 predicted that the number of transistors per silicon chip should double every two years. This prediction is considered by the semiconductor industries as Moore's law [1]. Figure 7.1 summarizes the history of microprocessor performance metrics [2]. Robert H. Dennard observed that with scaling, the clock frequency can be increased because delay scales and power density remain almost constant. Dennard's theory came to an end by the mid to late 2000 due to increase in OFF-state leakage current and consequently increased power dissipation to the point on which threshold and frequency scaling were no longer possible [3] and clock rate got settled near 4 GHz. This gave rise to multi-core processors.

DOI: 10.1201/9781003121589-7

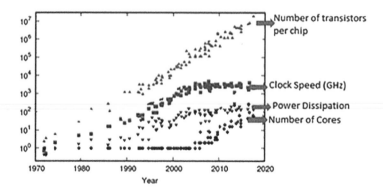

FIGURE 7.1   Industry trends for microprocessors over the last 42 years [2].

When a conventional MOSFET is scaled beyond 100 nm, various short channel effects (SCEs) become vital and affect the overall performance of the device. Due to the complexity of the fabrication and characterization processes, scaling beyond 100 nm becomes a very serious challenge for the VLSI industry. For energy efficient and high-speed performance, SCEs should be minimized. Further, to mitigate various problems associated with scaled technologies, a superior mechanism, and a process to control it, is needed. From the literature it is observed that steep sub-threshold slope devices that allow the device to operate at lower supply voltage, reduced DIBL, high I_{on}/I_{off} ratio, etc. are the best alternative for devices most susceptible to SCEs-related issues. International Technology Roadmap for Semiconductors (ITRS) [4] came up with novel MOSFET structures with better electrostatic gate control like silicon-on-insulator (SOI) devices, multi-gate MOSFET, FinFET devices, etc. Multi-gate devices and FinFET devices dominate the industry, as they provide better control of channel, reduced SCEs, improved sub-threshold slope, better I_{on}/I_{off}, and enhanced scalability.

The double-gate (DG) MOSFETs prove to be the potential candidate over the conventional MOSFET and single-gate SOI MOSFET, as they provide better control over the channel, higher ON-current, high ON current/OFF current, higher transconductance [5], steep sub-threshold slope, and suppress SCEs, thus enabling better switching characteristics [6].

7.2   CHOICE FOR DIELECTRIC MATERIAL: CONVENTIONAL SIO$_2$ VERSUS HIGH-*k* MATERIAL

7.2.1   SiO$_2$ as Dielectric

In conventional MOSFET, as shown in Figure 7.2, SiO$_2$ (silicon dioxide) is used as a gate dielectric. SiO$_2$ is a perfect choice as a dielectric for MOSFET for several years to prevent flow of current from the substrate to the gate terminal or vice versa, as it can withstand a high electric field (of order of MV/cm), thereby allowing thin layers. Other important features of SiO$_2$, such as low bulk fixed charge density, low interface state density, and manufacturability, have allowed the VLSI industry to keep SiO$_2$ as the best

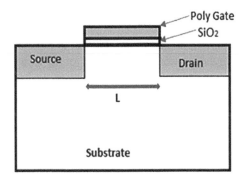

FIGURE 7.2 2D structure of conventional MOSFET.

option for MOSFET. But with the advancement in technology, oxide thickness keeps on reducing. At 20 nm gate length, SiO_2 is reduced to few atomic layers. This results in a high field causing a small tunneling current to flow from the gate to the body, and vice versa [7]. Figure 7.3a corresponds to the energy-band diagram in a MOSFET at flat-band condition with polysilicon as gate, SiO_2 as oxide, and p-type substrate. φ_o represents the barrier height at Si–SiO_2 interface. When V_G is positive as shown in Figure 7.3b, electrons tunnel from the substrate to the gate, giving rise to small tunneling current. Figure 7.3c depicts an energy band diagram when V_G is negative and electrons tunnel from the gate to the substrate, giving rise to small tunneling current [8].

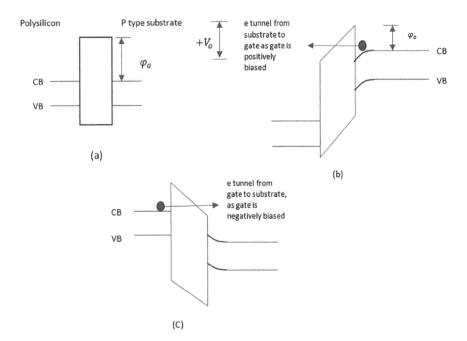

FIGURE 7.3 Energy band diagram of MOSFET showing tunneling of electrons (a) at flat-band condition, (b) when V_G is positive, and (c) when V_G is negative [8].

The tunneling of electrons can be guided by either the Fowler–Nordheim (FN) mechanism or by a direct tunneling (DT) mechanism, depending on the thickness and shape of potential barrier. FN tunneling mechanism is characterized by a triangular potential barrier and higher potential drop across oxide than the barrier height for electrons in the conduction band. Voltage drop across oxide, $V_o = V_{gs} - V_{fb} - \varphi_s - V_{poly}$. FN tunneling generally occurs for oxide greater than 4 nm thickness. The tunneling current density in FN mechanism can be calculated using [8]

$$J_{FN} = \frac{q^3 E_o^2}{16\pi^2 \hbar \varphi_o} \exp\left(-\frac{4\sqrt{2m^*}\varphi_o^{3/2}}{3\hbar q E_o}\right) \tag{7.1}$$

where q is charge of electron, E_o is the field across the oxide, φ_o is the barrier height for electrons, and m^* is the effective mass of an electron. For normal device operation, the FN tunneling current is negligible. Direct tunneling mechanism is characterized by trapezoidal potential barrier, thin oxide layer (generally less than 4 nm), and lower potential drop across oxide than the barrier height. The direct tunneling current density can be calculate using [9]

$$J_{DT} = \frac{q^3 E_o^2}{16\pi^2 \hbar \varphi_o} \exp\left(-\frac{\frac{4\sqrt{2m^*}\varphi_o^{3/2}}{3\hbar q}\left[1 - \left(1 - \frac{V_o}{\varphi_o}\right)^{\frac{3}{2}}\right]}{E_o}\right) \tag{7.2}$$

Gate leakage in direct tunneling for an n-type MOSFET as shown in Figure 7.3 can be due to tunneling of electrons from the conduction band of a p-type substrate to the conduction band of polysilicon and tunneling of holes from polysilicon to substrate [10].

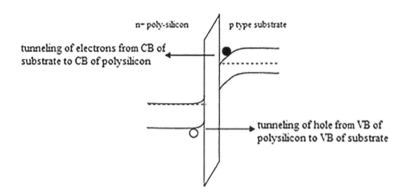

FIGURE 7.4   Mechanism of gate leakage in direct tunneling [11].

7.2.2 High-*k* Material as Dielectric

With the advancement of technology, as MOSFET is scaled down, the oxide thickness is also reduced, which results in increased tunneling current as discussed in the earlier section due to high field associated with thin oxide layers. The possible solution to reduce this increasing tunneling current is to use a thicker dielectric layer, but it would subsequently affect the driving capability of device by reducing the capacitance across the dielectric. The alternate solution is to use a dielectric material with high dielectric constant, or high *k*, allowing the material to store more charge. In other words, we can say, the coupling between two conducting plates becomes strong with high-*k* material and results in higher gate capacitance. Depending on the value of *k*, a dielectric material can be characterized as low-*k* material if *k* < 3.9 or high-*k* material if k > 3.9. The equivalent oxide thickness of a high-*k* material (EOT_{hk}) is given by [12, 13]

$$EOT_{hk} = \frac{\epsilon_{ox}}{\epsilon_{hk}} t_{hk} \tag{7.3}$$

$$t_{oxeff} = t_{ox} + EOT_{hk} \tag{7.4}$$

Where ϵ_{ox} is permittivity of the oxide layer, ϵ_{hk} is permittivity of the high-*k* dielectric material, t_{hk} is the thickness of the high-*k* layer, and t_{ox} is SiO_2 (interfacial) layer thickness. Application of high-*k* material results in reduced value of equivalent thickness, EOT_{hk}, as given by (7.3). The total effective oxide thickness is calculated as sum of interfacial layer thickness, t_{ox}, and equivalent thickness of a high-*k* material, EOT_{hk}, as given by (7.4) where t_{oxeff} is effective oxide thickness.

Various dielectric materials, their dielectric constant, their bandgaps, advantages and disadvantages are summarized in Table 7.1. To avoid tunneling, dielectric material must have bandgap greater than 1eV, and materials having bandgap greater than 5eV work perfectly, as they are able to induce more barrier heights thereby reducing the tunneling current. Depending on desired properties, the choice for high-*k* dielectrics includes La_2O_3, Al_2O_3, TiO_2, Ta_2O_5, HfO_2, ZrO_2, and Si_3N_4. Based on their properties we can conclude that Hafnium dioxide proves to be the best-suited high-*k* material for replacing SiO_2 and has been explored and used a lot in the literature. Table 7.2 summarizes the various methods used to synthesize HfO_2 thin films that have been explored in the literature, their advantages and disadvantages. The solgel method explained by Nishide in [38] is the best method adopted to synthesize HfO_2 high-quality thin films.

Although HfO_2 is a promising high-*k* material, its application in MOSFET has some drawbacks, such as the formation of voids and interlayer, mobility degradation and charge trapping, making a trade-off between reduction in tunneling current and mobility enhancement. Direct contact between HfO_2 and silicon substrate results in the formation of voids and an interlayer of hafnium silicate or silicide because of thermal instability of HfO_2. Dielectric constant of a high-*k* material is a result of higher ionic polarization, which causes phonon scattering and thereby

TABLE 7.1

Summary of Properties of Various Dielectric Materials [14–18]

High-*k* Dielectric Material	Dielectric Constant (*k*)	Band Gap [eV]	Advantages	Disadvantages	Reference
Al_2O_3	10	8.8	Large bandgap, Good thermal stability	Low dielectric constant	[19]
Si_3N_4	7.5	5	Good mechanical properties, Corrosion / thermal-shock resistance	Low dielectric constant	[20, 21]
La_2O_3	30	6	High dielectric constant, Thermodynamically stable with Si	Low bandgap	[22]
TiO_2	80	3.5	High dielectric constant	Unstable with silicon substrates	[23–25]
Ta_2O_5	22	4.4	High dielectric constant	Unstable with silicon substrates	[23–25]
ZrO_2	25	5.8	Excellent thermodynamic and chemical stability with Si	Less stable than HfO_2 with Si, leakage current is higher than that of HfO_2	[26–30]
HfO_2	20–25	5.5–6	More stable than ZrO_2 with the Si substrate, lower leakage current under the electric field 1 MV/cm, wider band gap, high dielectric constant, suitable band offset values relative to Si substrate, Excellent thermodynamic and chemical stability with Si	Formation of voids, mobility degradation	[26–30]

TABLE 7.2
Summary of Results of Different Coating Methods of HfO$_2$ [30]

Method Name	Process	Advantages	Disadvantages	Electrical properties
Chemical Vapor Deposition (CVD) [31]	Chemical reaction between gaseous reactant and heated surface	Flexible method, uniform films, good throwing power, low processing temperature, adjustable growth rate	High cost, difficulty in controlling the composition, sophisticated reactor composition impurity,	$k = 17$–23 EOT= 2.73 nm at 700°C
Atomic layer Deposition (ALD) [32]	substrate is exposed to alternating precursors layer by layer.	Sequential process, atomic level control, excellent step coverage, excellent conformal deposition, ultra-thin films, low processing temperature, excellent reproducibility, sharp edges can be produced	Very slow growth rate, confined by chamber size, residues remain in the chamber	$k = 12$–20 EOT = 0.5 nm at 800°C
Pulsed Laser Deposition (PLD) [33,34]	Target is illuminated by a focused pulsed laser beam	Preserve stoichiometric ratio, good adhesion, flexible, large deposition rate, low growth temperature, controlled deposition, does not need expensive precursor	Chunks, debris generation, defects in substrate because of large kinetic energy	$\kappa = 25$–40 EOT = 7.44 nm at 500°C
e-beam evaporation [35]	Target is bombarded with an electron beam	High purity, high material utilization efficiency, high deposition rates, freedom from wear	Low throughput, poor step coverage, filament degradation in the electron gun, higher surface roughness, relatively difficult to control film	$\kappa = 21$ EOT = 10.9 Å at 400°C
Molecular Beam Epitaxy [36]	Beam of molecules is bombarded on a heated surface	Low temperature processing, high purity, low defects, highly uniform	slow growth rate, not suitable for mass production, complex equipment	$\kappa = 17$–19 EOT = 1.3 nm at 500°C
RF sputtering [37]	Alternating current at radio frequency is required to avoid charge building up.	Better film quality, high step coverage, low pressure processing, no x-ray damage, lower contaminants, controllable growth,	Sputtered target plasma induced damage, difficult to deposited complex structure	$\kappa = 14$ EOT = 1.3 nm at 600°C
sol–gel [38]	A wet chemical technique that uses sol (a chemical solution) to produce an integrated network (gel)	high purity, low temperature processing, high adhesion, low cost, better homogeneity for multi-component materials, versatile, less energy consumption	Cost of precursor, fracture in gel due to capillary stresses, difficult to avoid residual porosity and OH groups.	$\kappa = 15$–22 EOT = 3.10 nm at 400°C

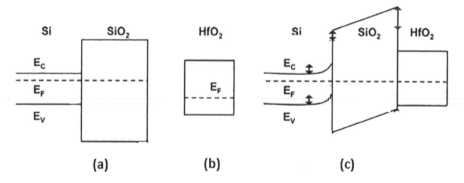

FIGURE 7.5 Energy band diagram at equilibrium of (a) SiO$_2$/Si stack, (b) HfO$_2$ alone, and (c) HfO$_2$/SiO$_2$/Si stack [39].

reducing the mobility of charge carriers in the channel [40]. High-*k* materials are also characterized by high interface traps, resulting in charge trapping. In order to overcome the above-mentioned drawbacks of direct contact of HfO$_2$/silicon substrate and simultaneously get the advantages of high-*k*, it is being suggested to grow an ultra-thin layer of SiO$_2$ between the high-*k* material and the Si substrate. Thus, HfO$_2$/SiO$_2$/Si ensures reduction in tunneling current along with mobility enhancement [49, 50].

When SiO$_2$ is brought in contact with Si substrate, their Fermi level aligns as shown in the Figure 7.5a. There is no transfer of electrons between SiO$_2$ and Si substrate, resulting in no band bending because of a wider band gap of SiO$_2$ (8.4 eV to 11eV). Fermi-level HfO$_2$ alone is lower than that of SiO$_2$/Si stack as shown in energy band diagram of HfO$_2$ (Figure 7.5b). When HfO$_2$ is brought in contact with SiO$_2$/Si stack, there is a transfer of electrons from SiO$_2$/Si stack to HfO$_2$, and their Fermi level aligns, resulting in an upward shift in the band at the Si interface as shown in Figure 7.5c.

7.3 CHOICE FOR GATE MATERIAL: POLYSILICON GATE VERSUS METAL GATE

Over the years, polysilicon was widely used as a gate electrode in MOSFET but with the advancement of technology, use of polysilicon as gate material is of great concern below 70 nm because of the challenges associated with polysilicon as gate electrode [41,42]. Especially while working with high-*k* dielectrics, selection of a gate electrode is another important issue along with selection of a high-*k* material as dielectric. Doped polysilicon lowers the MOS devices (mainly fabricated with high-*k* dielectric) gate capacitance because of the charge depletion effect. Stability of a gate electrode during high temperature processing (mainly during annealing) is of great concern. Metal gates having high work function are preferred for high-temperature CMOS processing. But these metal gates are generally unstable, as their work function changes during high-temperature processing. This instability can be overcome using the gate-last process in which process is first carried out in a conventional way with poly-Si, then poly-Si is removed, and the metal gate is deposited after high-temperature processing [43]. Metal gate is the preferred choice over poly-Si, as these are more effective in screening surface phonons [44] and offer reduced gate resistance and desirable work function settings.

7.4   SIMULATION OF DGFET WITH APPLICATION OF HIGH-*K* DIELECTRIC AND METAL GATE

The schematic of DG MOSFET with high-*k* and metal gate is shown in Figure 7.6. To obtain the results for a DG MOSFET simulation, we used the Lombardi mobility model to describe the carrier mobility in the MOSFET inversion layer. Figure 7.7 demonstrates the simulated structure with meshing designed using 2D Visual TCAD [46]. The geometrical and material parameters used are listed in Table 7.3.

The lateral Gaussian doping profile and mobility in the simulated structure are shown in Figure 7.8.

The effective/equivalent oxide thickness of HfO_2 *i.e.* EOT_{HfO2} can be calculate using equation (7.3) as follows:

$$EOT_{HfO_2} = \frac{\epsilon_{SiO_2}}{\epsilon_{HfO_2}} t_{HfO_2} = 0.49\,nm$$

$$t_{oxeff} = t_{SiO_2} + EOT_{HfO_2} = 1.1\,nm$$

where t_{oxeff} is total oxide thickness.

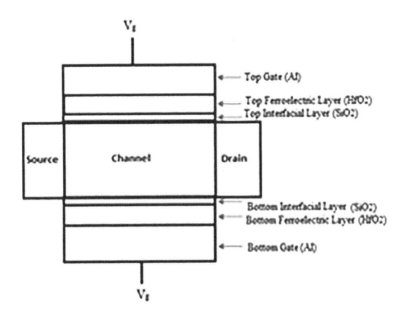

FIGURE 7.6   Schematic of a high-*k* gate stack DG MOSFET.

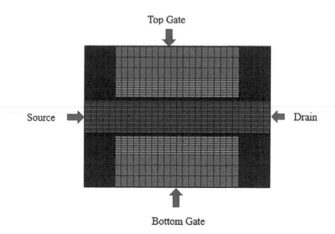

FIGURE 7.7   Simulated structure using Visual TCAD showing doping concentration and meshing.

A close comparison of transfer characteristics of the simulated structure (shown with solid line) with conventional DG MOSFET (shown with dotted line) with varying V_{ds} is shown in Figure 7.9. I_{ON} is improved with the application of high-k. From the figure it can be observed that I_{OFF} is reduced by large amount resulting in improved I_{ON} / I_{OFF} ratio. It should be worth noting that the simulated structure has higher threshold voltage than that of the conventional structure (without High-k) with improved I_{ON} / I_{OFF} ratio [45]. Thus, a conclusion can be made that the performance of conventional DG MOSFET is improved with the application of high-k dielectric resulting in steep subthreshold slope. Further it can be noticed that for low values of V_{ds} (50 mV), drain current is very low and it keeps on increasing with increase in V_{ds}. Figure 7.10 shows the comparison of $I_d - V_{ds}$ characteristics of high-k gate stack DG MOSFET (shown with solid line) with conventional DG MOSFET (shown with dotted line) with varying V_{gs}. With the introduction of High-k Gate Stack, ON current of the simulated structure is improved.

TABLE 7.3
Default Device Parameters

Parameter	Description	Value
L	Length of channel	20 nm
W	Width of channel	1 μm
t_{SiO2}	Interfacial oxide (SiO$_2$) thickness	0.6 nm
t_{HfO2}	High-*k* dielectric thickness	3 nm
t_m	Metal gate thickness	5 nm
t_{Si}	Si body thickness	5 nm
N_{SD}	Source/drain doping	1e+20
ϵ_{SiO2}	Interfacial oxide (SiO$_2$) permittivity	3.9
ϵ_{HfO2}	High-*k* dielectric permittivity	24
	Mobility model	Lombardi

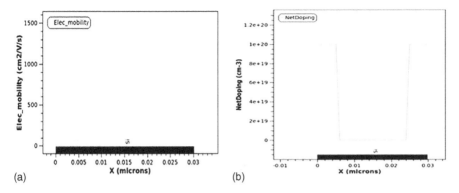

(a)   (b)

FIGURE 7.8 Simulated high-k gate stack DG MOSFET using Visual TCAD showing (a) lateral Gaussian doping profile and (b) mobility across the channel.

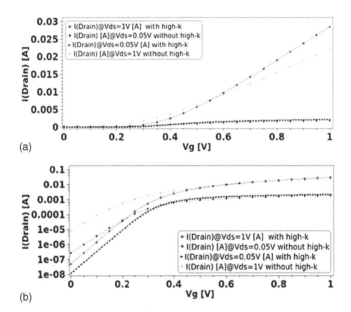

FIGURE 7.9 Comparison of transfer characteristics of simulated high-k gate stack DG MOSFET with conventional DG MOSFET with varying V_{ds} (a) in linear scale and (b) in logarithmic scale.

The impact of SiO_2 interfacial layer thickness is illustrated in Figure 7.11. Increased thickness of SiO_2 interfacial layer results in lowering of gate oxide capacitance, thereby reducing the drain current. Thinner interfacial layer results in higher ON-current and, therefore, higher transconductance. OFF-current is low with a thin interfacial layer, resulting in a steep sub-threshold slope. Thus, large drain current can be obtained with a thin interfacial layer because of high field associated with thin oxides. A comparative effect of varying thickness of high-k dielectric on transfer characteristics of DG MOSFET is shown in Figure 7.12. It can be observed from the

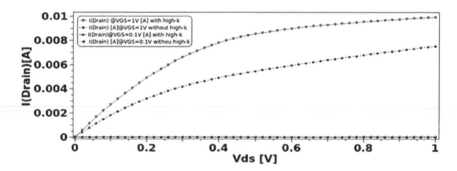

FIGURE 7.10 Comparison of $I_d - V_{ds}$ characteristics of high-*k* gate stack DG MOSFET with conventional DG MOSFET with varying V_{gs}.

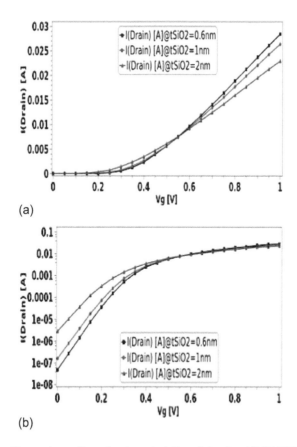

FIGURE 7.11 Comparison of transfer characteristics of simulated DG MOSFET with varying thickness of interfacial layer (a) in linear scale and (b) in logarithmic scale.

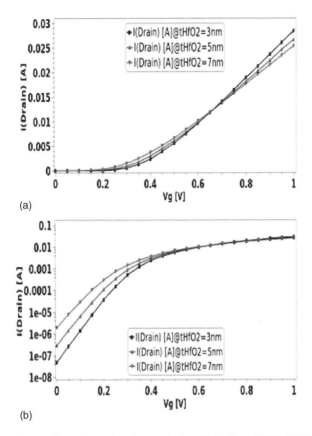

(a)

(b)

FIGURE 7.12 Comparison of transfer characteristics of simulated DG MOSFET with varying thickness of dielectric material (a) in linear scale and (b) in logarithmic scale.

transfer characteristics that C_{ox} is reduced with an increase in thickness of high-k dielectric, resulting in lowering of ON-current as seen in Figure 7.12a and the increase in OFF-current as seen in Figure 7.12b. The influence of the gate on the channel is dominant at 3 nm high-k thickness as compared to 7 nm. Thus, higher transconductance and higher current can be obtained with thin dielectric thickness.

The electrostatic potential across the channel is illustrated in Figure 7.13. The potential resembles the electrostatic potential obtained by 2D Poisson's Equation [47]:

$$\frac{\partial^2 \phi(x,y)}{\partial x^2} + \frac{\partial^2 \phi(x,y)}{\partial y^2} = \frac{qN_A}{\varepsilon_{Si}} \text{ for } 0 < y < t_{Si} \qquad (7.5)$$

The potential in the vertical direction can be approximated by a parabolic function from [48]:

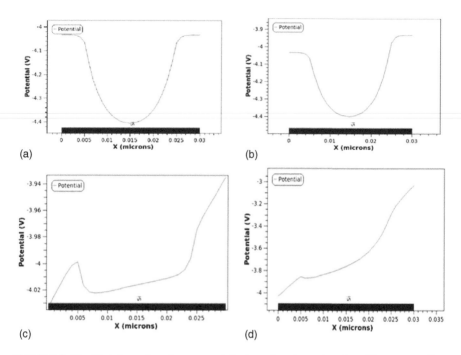

FIGURE 7.13 Variation of surface potential along the channel at L = 20 nm at $t_{Si}/2$ at (a) $V_{DS} = 0$ V and $V_{GS} = 0$ V; (b) $V_{DS} = 0.1$ V and $V_{GS} = 0$ V; (c) $V_{DS} = 0.1$ V and $V_{GS} = 1$ V; and (d) $V_{DS} = 1$ V and $V_{GS} = 0.8V$.

$$\phi(x,y) = \phi_s + a(x)y + b(x)y^2 \text{ for } 0 < y < t_{Si} \tag{7.6}$$

where ϕ_s is surface potential and a(x) and b(x) are coefficients calculated using the boundary conditions.

The surface potential at the source end is

$$\phi(x,y)\Big|_{x=0} = V_{bi} \tag{7.7}$$

The surface potential at the drain end is

$$\phi(x,y)\Big|_{x=L} = V_{bi} + V_{DS} \tag{7.8}$$

where V_{bi} is a built-in potential,

$$V_{bi} = \frac{E_g}{2q} + \phi_f \tag{7.9}$$

$$\phi_f = V_T \ln \frac{N_A}{n_i} \tag{7.10}$$

where E_g is silicon bandgap at 300 K, ϕ_f is Fermi potential, and V_T is thermal voltage.

The flat-band voltage, V_{FB}, at the gate terminal is given as

$$V_{FB} = \phi_M - \phi_S \tag{7.11}$$

where ϕ_M is the metal work function and ϕ_S is the silicon work function given by

$$\phi_S = \chi_{Si} + \frac{E_g}{2q} + \phi_f \tag{7.12}$$

Figure 7.13a shows the electrostatic potential variation at $V_{DS} = 0$ and $V_{GS} = 0$. The potential at source and drain end is equal to built-in potential. Because $V_{DS} = 0$, we note that the minimum potential position corresponding to the maximum barrier is located in the device center. As we increase the V_{DS} to 0.1V, the potential at source end is still equal to built-in potential, whereas the potential at drain end is increased to $V_{bi} + V_{DS}$ (as shown in Figure 7.13b), and the minimum potential position shifts toward the source. It can be observed that the energy barrier over which an electron at the source has to climb decreases with increasing V_{DS}. Figure 7.13c represents the electrostatic potential when we increase the gate voltage to 1 V (strong inversion). Higher gate bias results in an increase in surface potential, causing the minimum potential position to shift upward. Figure 7.13d represents the potential at drain voltage 1 V and gate potential 0.8 V.

Figure 7.14a–c illustrates the variation of surface potential with varying channel length. With an increase in the length of channel, the width of the barrier keeps increasing, thereby increasing the threshold voltage of the device. Figure 7.14d illustrates the potential variation from top gate to bottom gate. The potential becomes maximum at the center of the channel, and the electric field corresponding to this point becomes zero.

Electron density along the channel length is shown in Figure 7.15. We are discussing the effect of drain voltage on the channel when V_{GS} is far greater than V_t, which means a strong inversion had occurred and electrons are in the channel for the conduction. In Figure 7.15, a drain voltage is small, nearly 0.1 V, i.e., the width of the depletion region is almost same along the source and drain side, so electron density is almost the same at nearly $10^{19}/cm^3$. As we increase the drain voltage, more potential difference will occur between the drain and source terminal, so more electrons will inject in the channel at higher velocity, and consequently current increases linearly with drain voltage. If we further increase the drain voltage to 1 V, the drain terminal is more reverse biased, depletion region width is more at the drain side than the source side, the channel is tapered in the direction from the source to the drain, and pinch-off occurs. This means the electron density at the source end is

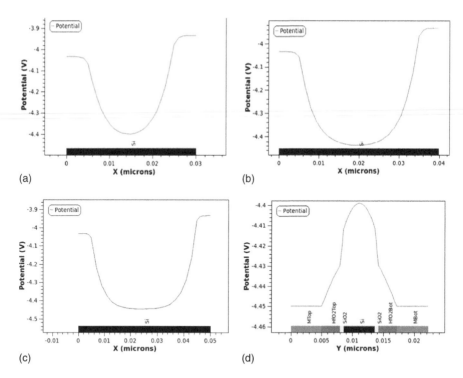

FIGURE 7.14 Surface potential variation along the channel length for different channel length with $V_{DS} = 0.1$ *V* and $V_{GS} = 0$ *V* at $t_{Si}/2$ at (a) L = 20 nm; (b) at L = 30 nm; (c) at L = 40 nm; and (d) top-to-bottom gate potential distribution.

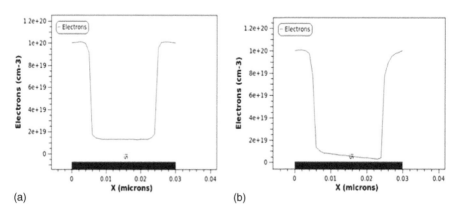

FIGURE 7.15 Electron density along the channel at $t_{Si}/2$ (a) at $V_{DS} = 0.1$ *V* and $V_{GS} = 1V$ and (b) at $V_{DS} = 1$ *V* and $V_{GS} = 0.8V$.

more than that at the drain terminal. Since pinch-off had occurred, current saturates. The tapered channel is clearly visible in Figure 7.15b. The effect of increasing drain voltage on the channel is also represented in terms of net charge density in Figure 7.16. The tapering of the channel with increasing drain voltage is clearly shown in Figure 7.16b.

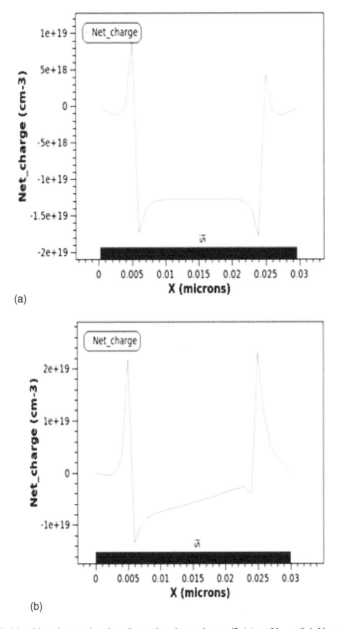

FIGURE 7.16 Net charge density along the channel at $t_S/2$ (a) at $V_{DS} = 0.1$ V and $V_{GS} = 1V$ and (b) at $V_{DS}1$ V and $V_{GS} = 0.8V$.

Figure 7.17 shows the variation of electric field along the channel for high-k DG MOSFETs. It is observed that the electric field in a high-k gate stack DG MOSFET (Figure 7.17b) is higher than in a DG MOSFET without high-k (Figure 7.17a). From the figure it is clear that the electric field at the drain side is slightly high, causing a more pronounced hot electron effect. But because of high-k, the electric field at the source

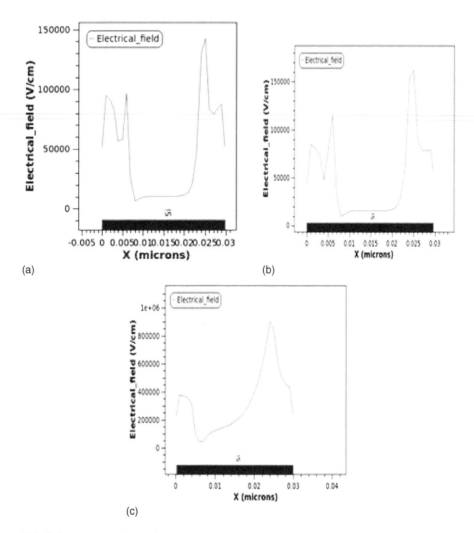

FIGURE 7.17   Variation of electric field along the channel at L = 20 nm at $t_{Si}/2$ (a) in DG MOSFET $V_{DS} = 0.1$ *V* and $V_{GS} = 1V$; (b) in high-*k* DG MOSFET at $V_{DS} = 0.1$ *V* and $V_{GS} = 1V$; and (c) in high-*k* DG MOSFET at $V_{DS} = 1$ *V* and $V_{GS} = 0.8V$.

end also increases, causing fast acceleration of carriers from source to drain, showing an increase in the electron injection velocity into the channel, and resulting in high speed operation. Thus, overall, the performance of a high-*k* gate stack DG MOSFET is improved. When drain-to-source voltage is increased further, electric field increases; this high field region expands toward the source with increasing drain voltage (Figure 7.17c); and the device behaves as the effective channel length has been reduced.

Figure 7.18 shows the effect of drain-to-source voltage and gate-to-source voltage on band bending. Figure 7.18a represents the energy band diagram when $V_{DS} = 0$ V

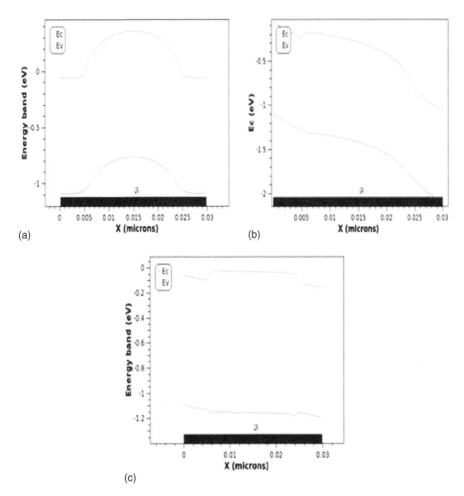

(a)

(b)

(c)

FIGURE 7.18 Effect of V_{DS} and V_{GS} on band bending in high-k gate stack DG MOSFET having L = 20 nm at $t_{Si}/2$ (a) at $V_{DS} = 0$ V and $V_{GS} = 0V$; (b) at $V_{DS} = 1$ V and $V_{GS} = 0.1V$; and (c) at $V_{DS} = 0.1$ V and $V_{GS} = 1V$.

and $V_{GS} = 0V$, and the electrons have a potential barrier or a potential hill that prevents them from flowing from the source to the drain. Figure 7.18b shows the effect of increasing drain-to-source voltage. When drain-to-source voltage is increased, the barrier height is reduced, thus enabling the source to inject carriers into the channel independent of gate voltage. This short channel effect is commonly known as drain-induced barrier lowering (DIBL) [7]. Figure 7.18c shows the effect of increasing gate-to-source voltage. Increasing gate voltage results in an increase in surface potential, causing the downward bending of the energy bands.

As seen from the transfer characteristics of a high-k gate stack DG MOSFET, the application of high-k dielectric results in improved overall performance of the device

TABLE 7.4

Comparison of I_{ON}/I_{OFF} with and without High-*k* Dielectric with Varying Channel Length at $V_{DS} = 1V$

Length of the Channel	DG-MOSFET without High-*k*			DG-MOSFET with High-*k*		
	I_{ON} (A)	I_{OFF} (A)	I_{ON}/I_{OFF}	I_{ON} (A)	I_{OFF} (A)	I_{ON}/I_{OFF}
L = 20 nm	0.0233	8.57e−6	2.72e+3	0.0285	4.583e−8	6.22e+5
L = 30 nm	0.0174	5.732e−8	3.04e+5	0.0227	3.783e−9	6.00e+6
L = 40 nm	0.0129	5.695e−9	2.26e+6	0.0205	1.292e−9	1.59e+7
L = 50 nm	0.0111	1.668e−9	6.64e+6	0.0178	6.683e−10	2.66e+7

with an increase in I_{ON} and a decrease in I_{OFF}. An increase in I_{ON} improves the driving capacity of the device. A decrease in I_{OFF} results in minimizing the static power consumption in standby mode, which is of great importance in today's era of battery-operated devices. Table 7.4 shows the comparison of I_{ON} / I_{OFF} with and without high-*k* with varying channel length at $V_{DS} = 1V$. From the table it is clearly visible that I_{ON} / I_{OFF}, which defines the device performance, improves. Figure 7.19 shows the comparison of variation in I_{ON} with an increase in channel length. I_{ON} decreases with an increase in channel length. Figure 7.20 shows a reduction in OFF-current with an increase in channel length in both high-*k* and no high-*k* DG MOSFET. Figure 7.21 depicts the comparison of the ON-current to OFF-current ratio with increasing channel length. The I_{ON} / I_{OFF} improves with the application of high-*k* dielectric.

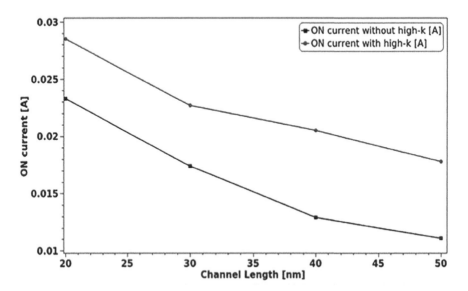

FIGURE 7.19 Comparison of variation in I_{ON} with physical channel length varying from 20 nm to 50 nm with and without high-*k* dielectric at $V_{DS} = 1V$.

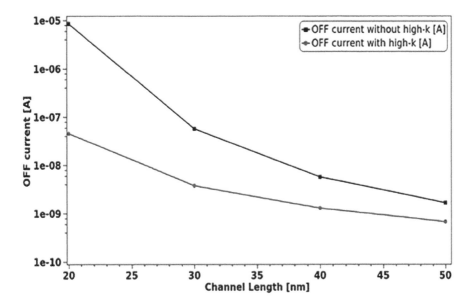

FIGURE 7.20   Comparison of variation in I_{OFF} with physical channel length varying from 20 nm to 50 nm with and without high-k dielectric at $V_{DS} = 1V$.

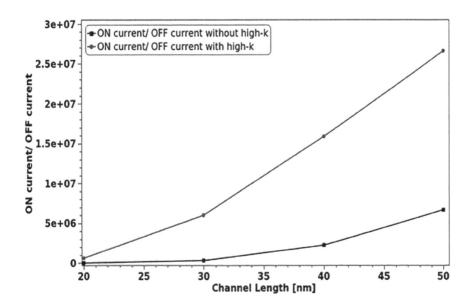

FIGURE 7.21   Comparison of variation in I_{ON} / I_{OFF} with and without high-k dielectric with varying channel lengths at $V_{DS} = 1V$.

Table 7.5 shows a comparison of a sub-threshold slope with and without high-*k* dielectric with varying channel lengths. It can be observed from the Figure 7.22 that a steep sub-threshold slope can be obtained using high-*k* dielectric.

Comparison of threshold voltage with and without high-*k* dielectric with varying channel lengths is shown in Table 7.6. With high-*k* dielectric, the threshold voltage is higher as compared to its counterpart without high-*k*. Variations of threshold voltage with varying physical lengths from 20 nm to 50 nm with and without high-*k* are shown in Figure 7.23.

TABLE 7.5

Comparison of Sub-threshold Slope with Physical Channel Length Varying from 20 nm to 50 nm with and without High-*k* Dielectric at $V_{DS} = 1V$

Length of the Channel	DG-MOSFET without High-*k* Dielectric (mV/dec)	DG-MOSFET with High-*k* Dielectric (mV/dec)
L = 20 nm	~80.4	~69
L = 30 nm	~72.3	~65.4
L = 40 nm	~68.1	~63.9
L = 50 nm	~64.9	~61.8

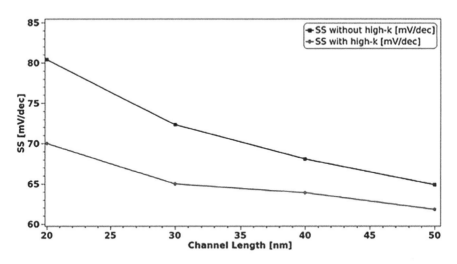

FIGURE 7.22 Comparison of sub-threshold slope with physical channel length varying from 20 nm to 50 nm with and without high-*k* dielectric at $V_{DS} = 1V$.

TABLE 7.6

Threshold Voltage for Physical Lengths 20 nm to 50 nm with and without High-*k* at $V_{DS} = 1V$

Length of the Channel	DG-MOSFET without High-*k*	DG-MOSFET with High-*k*
L = 20 nm	0.068	0.2028
L = 30 nm	0.1969	0.2515
L = 40 nm	0.2371	0.2701
L = 50 nm	0.2454	0.2794

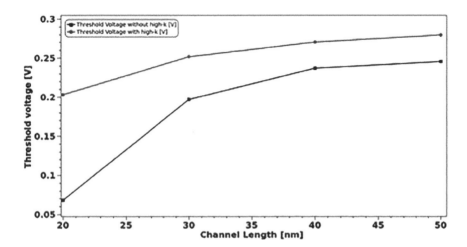

FIGURE 7.23 Comparison of variation of threshold voltage with varying physical lengths from 20 nm to 50 nm with and without high-*k* at $V_{DS} = 1V$.

7.5 CONCLUSION

With scaling, as the physical length of the channel is scaled down to a few nanometers, leakage current increases drastically, leading to high power consumption. With a high-*k* gate stack DG MOSFET, higher gate capacitance, increased speed, and reduced leakage current can be achieved. ON-current is also increased, resulting in a steep sub-threshold slope, which is very important for switching application. With application of a metal gate, low resistance and desirable work function can be achieved. Thus, a high-*k* DG MOSFET is a promising candidate for high speed and low power application.

REFERENCES

[1] G. E. Moore, "Cramming more components onto integrated circuits," *Electronics*, vol. 38, pp. 11–13, 1965.

[2] K. Rupp, "42 years of microprocessor trend data." Ed. by karlrupp.net. [Online; posted 15 February 2018]. February 2018. https://www.karlrupp.net/2018/02/42-years-of-microprocessor-trend-data/.

[3] M. H. Bohr, "A 30 Year Retrospective on Dennard's MOSFET Scaling Paper," *IEEE SSCS Newsletter*, vol. 12, no. 1, pp. 11–13, 2007.

[4] itrs2.netg, ed., "The International Technology Roadmap for Semiconductors." [Online; posted 15 May 2015]. May 2015. http://www.itrs2.net/itrs-reports.html.

[5] S. Tayal and A. Nandi, "Performance analysis of junctionless DGMOSFET based 6T SRAM with gate-stack configuration," *IET Micro & Nano Letters*, vol. 13, pp. 838–841, 2018.

[6] K. Koley, A. Dutta, B. Syamal, S. K. Saha, and C. K. Sarkar, "Subthreshold analog/RF performance enhancement of underlap DG FETs with high-K spacer for low power applications," *IEEE Transactions on Electron Devices*, vol. 60, no. 1, pp. 63–69, 2013, doi:10.1109/TED.2012.2226724.

[7] K. Roy, S. Mukhopadhyay, and H. Mahmoodi-Meimand, "Leakage current mechanisms and leakage reduction techniques in deep-submicrometer CMOS circuits," *Proceedings of the IEEE*, vol. 91, no. 2, pp. 305–327, 2003, doi:10.1109/JPROC.2002.808156.

[8] Y. Taur and T. H. Ning, *Fundamentals of Modern VLSI Devices*, New York: Cambridge University Press, 1998, pp. 95–97.

[9] K. Schuegraf and C. Hu, "Hole injection SiO_2 breakdown model for very low voltage lifetime extrapolation," *IEEE Transactions on Electron Devices*, vol. 41, pp. 761–767, 1994.

[10] K. M. Cao et al., "BSIM4 gate leakage model including source-drain partition," *International Electron Devices Meeting 2000*. Technical Digest. IEDM (Cat. No.00CH37138), San Francisco, CA, 2000, pp. 815–818, doi:10.1109/IEDM.2000.904442.

[11] A. Hokazono, K. Ohuchi, M. Takayanagi, Y. Watanabe, S. Magoshi, Y. Kato, T. Shimizu, S. Mori, H. Oguma, T. Sasaki, H. Yoshimura, K. Miyano, N. Yasutake, H. Suto, K. Adachi, H. Fukui, T. Watanabe, N. Tamaoki, Y. Toyoshima, et al., "14 nm gate length CMOSFETs utilizing low thermal budget process with poly-SiGe and Ni salicide," *IEEE International Electron Devices Meeting*, 2002, pp. 639–642.

[12] J. Robertson, "Band offsets of wide-band-gap oxides and implications for future electronic devices," *Journal of Vacuum Science and Technology B*, vol. 18, pp. 1785–1791, 2000.

[13] S. Tayal and A. Nandi, "Effect of FIBL in-conjunction with channel parameters on analog and RF FOM of FinFET," *Superlattice and Microstructures*, vol. 105, pp. 152–162, 2017.

[14] Q. Zeng, A. R. Oganov, A. O. Lyakhov, C. Xie, X. Zhang, J. Zhang, Q. Zhu, B. Wei, I. Grigorenko, L. Zhang, and L. Cheng, "Evolutionary search for new high-k dielectric materials: methodology and applications to hafnia-based oxides," *Acta Crystallographica*, vol. 70, p. 76, 2014.

[15] J. P. Locquet, C. Marchiori, M. Sousa, J. Fompeyrine, and J. Seo, "High-*k* dielectrics for the gate stack," *Journal of Applied Physics*, vol. 100, p. 051610, 2006.

[16] J. C. Pravin, D. Nirmal, P. Prajoon, and J. Ajayan, "Implementation of nanoscale circuits using dual metal gate engineered nanowire MOSFET with high-k dielectrics for low power applications," *Physica E*, vol. 83, p. 95, 2016.

[17] X. B. Lu, G. H. Shi, J. F. Webb, and Z. G. Liu, "Dielectric properties of $SrZrO_3$ thin films prepared by pulsed laser deposition", *Applied Physics A*, vol. 77, p. 481, 2003.

[18] W. Chen, W. Ren, Y. Zhang, M. Liu, and Z. G. Ye, "Preparation and properties of ZrO2 and TiO2 films and their nanolaminates by atomic layer deposition," *Ceramics International*, vol. 41, p. S278, 2015.

[19] S. J. Lee and S. Baek, "Effect of SiO_2 Content on the Microstructure, Mechanical and Dielectric Properties of Si_3N_4 Ceramics," *Ceramics International*, vol. 42, p. 9921, 2016.

[20] Y. Xia, Y. P. Zeng, and D. Jiang, "Dielectric and mechanical properties of porous Si_3N_4 ceramics prepared via low temperature sintering," *Ceramics International*, vol. 35, p. 1699, 2009.

[21] J. B. Cheng, A. D. Li, Q. Y. Shao, H. Q. Ling, D. Wu, Y. Wang, Y. Bao, M. Wang, G. Z. Liu, and N. B. Ming, "Growth and characteristics of La_2O_3 gate dielectric prepared by low pressure metalorganic chemical vapor deposition," *Applied Surface Science*, vol. 233, p. 91, 2004.

[22] M. K. Bera and C. K. Maiti, "Electrical properties of SiO_2/TiO_2 high-*k* gate dielectric stack," *Materials Science in Semiconductor Processing*, vol. 9, p. 909, 2006.

[23] C. E. Kim and I. Yun, "Effects of the interfacial layer on electrical characteristics of $Al_2O_3/TiO_2/Al_2O_3$ thin films for gate dielectrics," *Applied Surface Science*, vol. 258, p. 3089, 2012.

[24] M. Houssa, L. Pantisano, L. Å. Ragnarsson, R. Degraeve, T. Schram, G. Pourtois, S. D. Gendt, G. Groeseneken, and M. M. Heyns, "Electrical properties of high-κ gate dielectrics: Challenges, current issues, and possible solutions," *Materials Science and Engineering R*, vol. 51, p. 37, 2006.

[25] A. Salaün, H. Grampeix, J. Buckley, C. Mannequin, C. Vallée, P. Gonon, S. Jeannot, C. Gaumer, M. Gros-Jean, and V. Jousseaume, "Investigation of HfO_2 and ZrO_2 for Resistive Random Access Memory applications," *Thin Solid Films*, vol. 525, p. 20, 2012.

[26] W. Zheng, K. H. Bowen, J. Li, I. Dabkowska, and M. Gutowski, "Electronic Structure Differences in ZrO_2 vs HfO_2," *The Journal of Physical Chemistry A*, vol. 109, p. 11521, 2005.

[27] H. Shimizu and T. Nishide, *Advances in Crystallization Processes*, Ed. Y. Mastai, InTech, Rijeka, 2012, Ch. 13.

[28] P. Jin, G. He, D. Xiao, J. Gao, M. Liu, J. Lv, Y. Liu, M. Zhang, P. Wang, and Z. Sun, "Microstructure, optical, electrical properties and leakage current transport mechanism of sol-gel-processed high-k HfO_2 gate dielectrics," *Ceramics International*, vol. 42, p. 6761, 2016.

[29] S. S. Jiang, G. He, J. Gao, D. Q. Xiao, P. Jin, W. D. Li, J. G. Lv, M. Liu, Y. M. Liu, Z. Q. Sun, "Microstructure, optical and electrical properties of sputtered HfTiO high-k gate dielectric thin films," *Ceramics International*, vol. 42, p. 11640, 2016.

[30] S. Kol and A. Oral, "Hf-based high-κ dielectrics: A review," *Acta Physica Polonica A*, vol. 136, 2019. doi:10.12693/APhysPolA.136.873.

[31] K. J. Choi, W. C. Shin, J. B. Park, and S. G. Yoon, "Electrical characteristics and thermal stability of $Pt/HfO_2/Si$ metal-insulator-semiconductor capacitors," *Integrated Ferroelectrics*, vol. 48, p. 13, 2002.

[32] R. Gupta, R. Rajput, R. Prasher, and R. Vaid, "Structural and electrical characteristics of ALD- HfO_2/n-Si gate stack with SiON interfacial layer for advanced CMOS technology," *Solid State Sciences*, vol. 59, p. 7, 2016.

[33] H. Ikeda, S. Goto, K. Honda, M. Sakashita, A. Sakai, S. Zaima, and Y. Yasuda, "Local leakage current of HfO_2 thin films characterized by conducting atomic force microscopy," *Japanese Journal of Applied Physics*, vol. 41, p. 2476, 2002.

[34] G. He, L.Q. Zhu, M. Liu, Q. Fang, and L. D. Zhang, "Optical and electrical properties of plasma-oxidation derived HfO₂ gate dielectric films," *Applied Surface Science*, vol. 253, p. 3413 2007.

[35] Y. Wang, Z. Lin, X. Cheng, H. Xiao, F. Zhang, and S. Zou, "Study of HfO₂ thin films prepared by electron beam evaporation," *Applied Surface Science*, vol. 228, p. 93 2004.

[36] Y. K. Lu, X. F. Chen, W. Zhu, and R. Gopalkrishnan, "Growth and characterization of HfO₂ high-*k* gate dielectric films by laser molecular beam epitaxy (LMBE)," *Journal of Materials Science: Materials in Electronics*, vol. 17, p. 685, 2006.

[37] A. Vinod, M. S. Rathore, and N. S. Rao, "Effects of annealing on quality and stoichiometry of HfO₂ thin films grown by RF magnetron sputtering," *Vacuum*, vol. 155, p. 339, 2018.

[38] T. Nishide, S. Honda, M. Matsuura, M. Ide, "Surface, structural and optical properties of sol-gel derived HfO₂ films," *Thin Solid Films*, vol. 371, no. 1–2, pp. 61–65, 2000, https://doi.org/10.1016/S0040-6090(00)01010-5.

[39] X. Wang, H. Kai, W. Wang, J. Xiang, H. Yang, J. Zhang, X. Ma, C. Zhao, D. Chen, and T. Ye, "Band alignment of HfO2 on SiO2/Si structure," *Applied Physics Letters*, vol. 100, 2012, doi:10.1063/1.3694274.

[40] M. V. Fischetti, D. A. Neumayer, and E. A. Cartier, "Effective electron mobility in Si inversion layers in metal–oxide–semiconductor systems with a high-κ insulator: The role of remote phonon scattering," *Journal of Applied Physics*, vol. 90, p. 4587, 2001, https://doi.org/10.1063/1.1405826

[41] S. Q. Wang and J.W. Mayer, "Reactions of Zr thin films with SiO₂ substrates," *Journal of Applied Physics*, vol. 64, pp. 4711–4716, 1988.

[42] S. Tayal and A. Nandi, "Optimization of gate-stack in junctionless Si-nanotube FET for analog/RF performance," *Materials Science in Semiconductor Processing*, vol. 80, pp. 63–67, 2018.

[43] K. Mistry, C. Allen, C. Auth, B. Beattie, D. Bergstrom, M. Bost, M. Brazier, M. Buehler, A. Cappellani, R. Chau, C.H. Choi, G. Ding, K. Fischer, T. Ghani, R. Grover, W. Han, D. Hanken, M. Hattendorf, J. He, et al., "A 45 nm logic technology with high-k+ metal gate transistors, strained silicon, 9 Cu interconnect layers, 193 nm dry patterning, and 100% Pb-free packaging," *IEEE International Electron Devices Meeting*, 2007, pp. 247–250.

[44] S. Tayal and A. Nandi, "Enhancing the delay performance of junctionless Si nanotube based 6T SRAM," *IET Micro & Nano Letters*, vol. 13, pp. 965–968, 2018.

[45] E. Goel, S. Kumar, G. Rawat, M. Kumar, S. Dubey, and S. Jit, "Two dimensional model for threshold voltage roll-off of short channel high-k gate-stack double-gate (DG) MOSFETs," pp. 193–196, 2014. doi:10.1007/978-3-319-03002-9_48.

[46] Genius Semiconductor Device Simulator.

[47] K. K. Young, "Short-channel effect in fully depleted SOI MOSFET's," *IEEE Transactions on Electron Devices*, vol. 36, no. 2, pp. 399–402, 1989.

[48] G. V. Reddy and M. J. Kumar, "A new dual material double gate (DMDG) nanoscale SOI MOSFET: two dimensional analytical modeling and simulation," *IEEE Transactions on Electron Devices*, vol. 4, no. 2, pp. 260–268, 2005.

[49] S. Tayal and A. Nandi, "Analog/RF performance analysis of channel engineered high-K gate-stack based junctionless Trigate-FinFET," *Superlattices and Microstructures*, vol. 112, pp. 287–295, 2017, https://doi.org/10.1016/j.spmi.2017.09.031.

[50] S. Tayal and A. Nandi, "Interfacial layer dependence of High-K gate stack based Conventional trigate FinFET concerning analog/RF performance," *2018 4th International Conference on Devices, Circuits and Systems (ICDCS)*, Coimbatore, India, 2018, pp. 305–308, doi:10.1109/ICDCSyst.2018.8605172.

8 Novel Architecture in Gate-All-Around (GAA) MOSFET with High-*k* Dielectric for Biomolecule Detection

Krutideepa Bhol
VIT-AP University, Amaravati, India

Biswajit Jena
Koneru Lakshmaiah Education Foundation, Vaddeswaram, India

Umakanta Nanda
VIT-AP University, Amaravati, India

Shubham Tayal
SR University, Warangal, India

Amit Kumar Jain
University of Cambridge, Cambridge, UK

CONTENTS

8.1 INTRODUCTION

With the ever-increasing need for high-speed and low-power-consumption effective logic performances accompanied by low production costs of integrated circuits (ICs), fast development in the semiconductor industry is encouraged through continuous downscaling of transistor device dimensions. This perception of reduction shows a dynamic role in modifying the overall performance of the devices to attain the required incremented efficiency. Still, it has been noticed that as the dimensions are reduced to the nano-meter scale, the conventional planar metal oxide semiconductor field effect transistor (MOSFET) is no longer able to support Moore's law. Numerous physical boundaries related to the standard downscaling are associated with dependability problems such as increased gate leakage current, unwanted short channel effects, higher power consumption, degradation of the sub-threshold slope, etc. [1, 2]. Therefore, an alternative method to overcome the basic difficulties of scalability requires the implementation of advanced technology or execution of some latest planning resolving the power issues and short channel problems successively.

According to current studies the device performance can be improved through advanced gate electrostatic controllability to reduce the short channel effects. Additional advanced development is aided by implementing 1D channel geometries containing silicon nanowires, which, as covered in the gate-all-around (GAA) method, achieve the greatest controllability over the device channel, showing the best I_{on}/I_{off} ratio in comparison to the conventional planar and other multi-gate complements [3, 4].

Due to its attractive geometry and physical characteristic, it is likely to moderate the device ideas using the ultrathin nanowire as the upcoming generation of semiconductor devices. One such possible device stage is silicon nanowire, named reconfigurable nanowire gate-all-around metal oxide semiconductor field effect transistor (GAA-MOSFET), with unique reconfigurable properties of implementing n-type and p-type characteristics depending on the given bias polarization [5]. Some GAA-MOSFETs are produced with multiple-gate topologies as driver and control gates for application of controlling and drain current switching and polarization of the device [6].

Subsequently, a GAA-MOSFET can be accomplished with the nanowire structure; this reconfigurable device performance can be executed in the design of complex circuits to attain reprogrammable and complimentary CMOS operations with a smaller number of transistors. Therefore, an effective replacement to the conventional trend of nano-scaling is achieved by nanowire-based GAA-MOSFETs that operate multi-functionalities in a single-chip device by implementing each of n-type or p-type behavior by applied different biasing. This helps in enhancing the total number of functions per transistor without a decrease in the device geometry, consequently contributing advanced high integrated package densities at minimum device cost [7, 8].

8.2   DEVICE DESIGN AND SIMULATION

Three-dimensional schematic representation of the proposed structure with all the regions is indicated in Figure 8.1. The unique architecture with multiple nanowires forming the channel is represented in Figure 8.2 with channel numbers. In order to use it as a biosensor, the metal gate is removed either from the source side, the drain side, or both. These regions are named R1 and R2, as shown in Figure 8.3. A thin layer of Si_3N_4 is covering the individual nanowire followed by a thick layer of HfO_2 covering the whole channel. The gate metal chosen is titanium with work function of 4.6 eV. The cavity length under gate metal is represented as Lc. The simulation is carried out using Sentaurus TCAD. Sentaurus TCAD includes almost all the state-of-the-art device physics and simulation facilities for quantum potential–based devices also. This simulator is having high-standard GUI with both 2D and 3D design features.

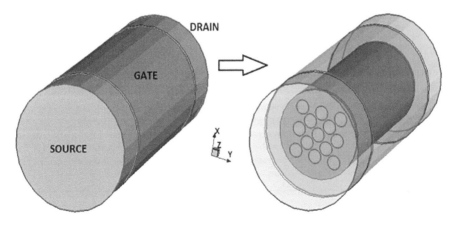

FIGURE 8.1   Dimensional device structure indicating different regions.

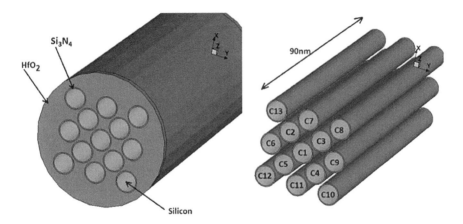

FIGURE 8.2   GAA MOSFET with individual nanowires covered with high-k dielectrics.

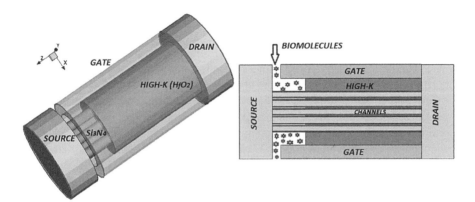

FIGURE 8.3   3D structure of (a) a high-k dielectric–based GAA biosensor and (b) a cut section indicating all the parts.

TABLE 8.1

Device Dimensions of the Proposed Structures

Device Dimensions	Nanowire Array GAA MOSFET	Nanowire Array GAA MOSFET with Cavity (as Biosensor)
Channel Length (L_g)	30–90 nm	30–90 nm
Doping Concentration(N_S/N_D)	Phosphorus (1e+20)	Phosphorus (1e+20)
Doping Concentration(N_a/N_d)	Boron (1e+16)	Boron (1e+16)
Oxide Thickness (t_{ox})	(1 nm Si_3N_4), (20 nm HfO_2)	(1 nm Si_3N_4), (20 nm HfO_2)
Channel Thickness (t_{si})	5 nm each (13 channels)	5 nm each (13 channels)
Source/Drain Extensions(L_s/L_d)	20 nm	20 nm

This simulator includes all the possibilities, not only for simple structures but also for more complex design and simulation to study the characteristics extensively. This simulator also includes different carrier transport models, such as drift diffusion model, hydrodynamic model, thermodynamic model, and quantum model, to study the movement of electrons inside the channel. The various parameters used in the simulation are tabulated in Table 8.1.

At the nano-scale level, the short channel effects (SCEs) on the characteristics of nanowire-based MOSFETs cannot be ignored. In order to reduce this SCEs impact, a higher number of multi-gate devices have been suggested, leading to the development of double-gate (DG) MOSFET, Fin-FETs, tri-gate MOSFET, and gate-all-around (GAA) MOSFET or surrounding-gate (SG) MOSFET, which can reduce the SCEs and enhance gate controllability [9,10]. Also at the nano-scale level, understanding of an ultra-sharp doping concentration profile between an n-type or p-type drain/source (D/S) region and an n-type or p-type body region still delivers a prominent difficulty in the fabrication process of nanowire short channel devices. Many new types of MOSFET designs have been suggested to overcome this hazardous technological problem. To date, one of the most fascinating developments in the nanotechnology area is the gate-all-around metal oxide semiconductor field effect transistor (GAA-MOSFET), which has received special consideration because of its outstanding electrical properties and large ability of protection from SCEs, producing low DIBL and improved I_{on}/I_{off} ratio [11, 12]. Colinge et al. [13] outlined a number of advantages of GAA-MOSFET, such as low DIBL low gate leakage current and simpler fabrication process than that of a conventional planer MOSFET [14, 15]. GAA-MOSFET is heavily doped with a silicon layer and has a different type of doping concentration in drain, source, and channel regions [16, 17].

Current literature also proposes that a nanowire GAA-MOSFET device beats the practical efficiency of conventional CMOS device in terms of leakage current, device performance, power, and area consumption. Furthermore, the exceptional feature of reconfigurability encourages protection of low power hardware devices through new nano-scaling devices [18]. Therefore, nanowire-based GAA-MOSFET structures with their variety of functional benefits are causing an outstanding level of research interest to discover their impact on various applications in the semiconductor device field [19].

A biosensor is a systematic device used for detection of biological components relating with the superb sensitivity and binding of biomolecules like antibodies, proteins, nucleotides, and enzymes. The biosensor has been used for various recent applications such as medical diagnosis [20], food analysis [21], drug development [22], environmental field monitoring [23], study of biomolecules and their interactions [24], crime detection [25], etc. Recently MOSFET-based biosensor devices have been considered very extensively due to their several benefits over other application methods with high electrostatic control, high sensitivity, direct transduction, mass production, and capability to reduce the short channel effects expressively. In the biosensor regime, they are also continuously gaining wide-ranging attention for the label-free discovery of neutral and charged biomolecules due to their higher sensitivity, miniaturization, and large probability with standard fabrication technology of CMOS [26]. In this nanowire-based GAA-MOSFET, the perpendicular nano-gaps are fixed in the gate material for binding of the biomolecules, which then cooperate with the device to achieve the modification of device electrostatics controllability and measure the sensitivity consequently. Still, structural unpredictability, high cost, and low binding risk are repeatedly restricting the application of biosensor devices [27, 28]. Hence, they are attaining the important attention as a possible alternative to GAA-MOSFET-based biosensors where gate underlap cavity is generated by etching both the gate dielectric and the gate material [29].

Current semiconductor device research studies describe the investigative models of nanowire implanted biosensors where all the biotargets are restricted within the fixed underlap nano-scaled region. The nanowire-based GAA-MOSFETs are widely used as biosensors with drain current variation potential swing and threshold voltage difference as biosensing measurement [30, 31]. In this chapter, the biosensing application of nanowire GAA-MOSFET for the discovery of biomolecules is explored with a minor structural adjustment. An investigative model of nanowire-based GAA-MOSFET for the biosensor application is developed with proper device simulation, appropriate nano-scaling measurements, and reducing the short channel effects for making a more accurate device study. The biomolecules then collaborate with the biosensor, the flat band potential is modified by distressing the threshold voltage, and surface potential of the device is considered for approximating the advance characteristic trend of the sensor device and its sensitivity parameter.

8.3 RESULTS AND DISCUSSIONS

Figure 8.4 indicates the electrostatic potential as well as electric field distribution in case of a nanowire array–based GAA MOSFET with a lower gate bias of 0.2 V and a drain bias of 0.5 V. The distribution of electrostatic potential from the source side to the drain side through the channel can be observed in Figure 8.4a. The distribution of electric field throughout the channel and dielectric can be observed in Figure 8.4b.

Figure 8.5 illustrates the potential and electric field distribution when a cavity is created below the gate metal for biomolecules detection. When the biomolecule comes in contact with the cavity, the potential changes, which allows the molecule to

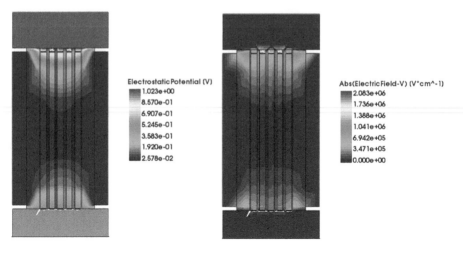

FIGURE 8.4   (a) Electrostatic potential and (b) electric field of high-*k* dielectric–based GAA MOSFET (without cavity).

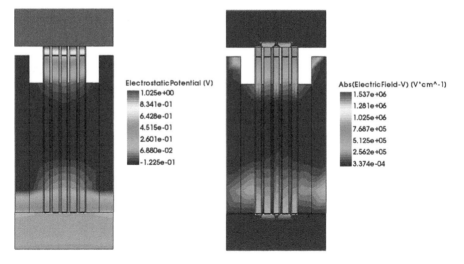

FIGURE 8.5   (a) Electrostatic potential and (b) electric field of high-*k* dielectric–based GAA MOSFET as biosensor.

be identified. At the same time the effect of biomolecules in the cavity can be identified by the electric field distribution. The corresponding values in a particular region can be observed from the legend associated with each cut section.

Figure 8.6 shows the mobility of electrons and holes inside the device and through silicon when a gate bias is applied. Apart from that, when a biomolecule enters the cavity and changes the potential of the device by its presence, the mobility also gets affected. The distributed electron and hole mobility can be easily realized from the figures to get an insight into the effect of biomolecules on device performance.

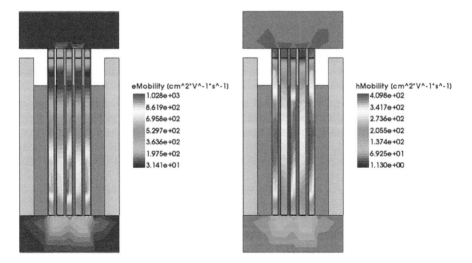

FIGURE 8.6   (a) eMobility and (b) hMobility of high-*k* dielectric–based GAA MOSFET as biosensor.

8.4   CONCLUSION

In this work, a novel architecture was designed to use nanowire GAA MOSFET as biosensor. With the gate-all-around structure, it is considered the best device for mitigating electrostatic controllability issues. At the same time, introduction of multiple nanowire as channel increases the drain current of the device compared to other device counterparts. Including high-*k* dielectric as a dielectric material, the device scaling entered into a lower technology node and further scaling down to use in biomedical applications. With improved results, the proposed structure can be a state-of-the-art counterpart to the devices dominating the semiconductor industry today.

REFERENCES

[1] Rechem D and Latreche S. The effect of short channel on nanoscale SOI MOSFETs. *Afr Rev Phys* 2008;2(38):80–81.
[2] D'Agostino F and Quercia D. Introduction to VLSI Design (EECS 467), Short-Channel Effects in MOSFETs, December 11th (2000).
[3] Tayal S and Nandi A. Optimization of gate-stack in junctionless Si-nanotube FET for analog/RF performance. *Mater Sci Semicond Process* 2018;80:63–67.
[4] Verhulst AS, Sorée B, Leonelli D, Vandenberghe WG, and Groeseneken G. Modeling the single-gate, double-gate, and gate-all-around tunnel field-effect transistor. *J Appl Phys* 2010;107:024518.
[5] Darbandy G, Claus M, and Schroter M. High-performance reconfigurable Si nanowire field-effect transistor based on simplified device design. *IEEE Trans Nanotechnol* 2016;15(2):289–294.
[6] Weber WM, Heinzig A, Trommer J, Martin D, Grube M, Mikolajick D. Reconfigurable nanowire electronic—A review. *Solid-State Electron* 2014;102:12–24.
[7] Tayal S and Nandi A. Effect of FIBL in-conjunction with channel parameters on analog and RF FOM of FinFET. *Superlattice Microst* 2017;105:152–162.

[8] Martin D, et al. Direct probing of Schottky barriers in Si nanowire Schottky barrier field effect transistors. *Phys Rev Lett* 2011;107(21):216807-1–216807-5.

[9] Park J-T and Colinge J. Multiple-gate SOI MOSFETs: Device design guidelines. *Electron Devices IEEE Trans* 2002;49(12):2222–2229.

[10] Tayal S and Nandi A. Analog/RF performance analysis of inner gate engineered junctionless Si nanotube. *Superlattice Microstr* 2017;111:862–871.

[11] Jin X, Liu X, Wu M, Chuai R, Lee J-H, and Lee J-H. A unified analytical continuous current model applicable to accumulation mode (junctionless) and inversion mode MOSFETs with symmetric and asymmetric double-gate structures. *Solid-State Electron* 2013;79:206–209.

[12] Jazaeri F, Barbut L, Koukab A, and Sallese J-M. Analytical model for ultra-thin body junctionless symmetric double gate MOSFETs in subthreshold regime. *Solid-State Electron* 2013;82:103–110.

[13] Lee C-W, Afzalian A, Akhavan ND, Yan R, Ferain I, and Colinge J-P. Junctionless multigate field-effect transistor. *Appl Phys Lett* 2009;94:053511.

[14] Lee C-W, Borne A, Ferain I, Afzalian A, Yan R, Akhavan ND, Razavi P, and Colinge J. High-temperature performance of silicon junctionless MOSFETs. *Electron Devices IEEE Trans* 2010;57:620–625.

[15] Yeh M-S, Wu Y-C, Wu M-H, Chung M-H, Jhan Y-R, and Hung M-F. Characterizing the electrical properties of a novel junctionless poly-Si ultrathin-body field-effect transistor using a trench structure. *IEEE Trans Electron Dev* 2015;36:150–152.

[16] Tayal S and Nandi A. Performance analysis of junctionless DGMOSFET based 6T SRAM with gate-stack configuration. *IET Micro Nano Lett* 2018;13:838–841.

[17] Sahu C and Singh J. Charge-plasma based process variation immune junctionless transistor. *IEEE Trans Electron Dev* 2014;35:411–413.

[18] Chen A, Hu XS, Jin Y, Niemier M, and Yin X. *Using emerging technologies for hardware security beyond PUFs. 2016 Design, Automation & Test in Europe Conference & Exhibition*, 2016. pp. 1544–1549.

[19] Heinzig A, Slesazeck S, Kreupl F, Mikolajick T, and Weber WM. Reconfigurable silicon nanowire transistors. *Nano Lett* 2012;12(1):119–124.

[20] Ribaut C, Loyez M, Larrieu JC, Chevineau S, Lambert P, Remmelink M, Wattiez R, and Caucheteur C. Cancer biomarker sensing using packaged plasmonic optical fiber gratings: Towards in vivo diagnosis. *Biosens Bioelectron* 2017;92:449–456.

[21] Sheikhshoaie M, Hassan KM, Sheikhshoaie I, and Ranjbar M. Voltammetric amplified sensor employing RuO_2 nano-road and room temperature ionic liquid for amaranth analysis in food samples. *J Mol Liq* 2017;229:489–494.

[22] Stobiecka M, Jakiela S, Chalupa A, Bednarczyk P, and Dworakowska B. Mitochondria–based biosensors with piezometric and RELS transduction for potassium uptake and release investigations. *Biosens Bioelectron* 2017;88:114–121.

[23] Nguyen VT, Kwon YS, and Gu MB. Aptamer-based environmental biosensors for small molecule contaminants. *Curr Opin Biotechnol* 2017;45:15–23.

[24] Makwana AB, Darjee S, Jain VK, Kongor A, Sindhav G, and Rao MV. A comparative study: Metal nanoparticles as fluorescent sensors for biomolecules and their biomedical application. *Sens Actuators Chem* 2017;246:686–695.

[25] Shin J, Choi S, Yang JS, Song J, Choi JS, and Jung HI. Smart forensic phone: Colorimetric analysis of a bloodstain for age estimation using a smartphone. *Sens Actuators B Chem* 2017;243:221–225.

[26] Buvaneswari B and Balamurugan NB. 2D analytical modeling and simulation of dual material DG MOSFET for biosensing application. *Int J Electron Commun (AEÜ)* 2019;99:193–200.

[27] Im H, Huang X-J, Gu B, and Choi Y-K. A dielectric-modulated field-effect transistor for biosensing. *Nature Nanotechnol* 2007;2(7):430–434.

[28] Gu B, Park TJ, Ahn J-H, Huang X-J, Lee SY, and Choi Y-K. Nanogap field-effect transistor biosensors for electrical detection of avian influenza. *Small* 2009;5(21):2407–2412.

[29] Kim J-Y, Ahn J-H, Moon D-I, Park TJ, Lee SY, and Choi Y-K. Multiplex electrical detection of avian influenza and human immunodeficiency virus with an underlapembedded silicon nanowire field-effect transistor. *Biosens Bioelectron* 2014;55:162–167.

[30] Ajay A, Narang R, Saxena M, and Gupta M. Drain current model of a fourgate dielectric modulated MOSFET for application as a biosensor. *IEEE Trans Electron Devices* 2015;62(8):2636–2644.

[31] Kinsella JM and Ivanisevic A. Biosensing: taking charge of biomolecules. *Nature Nanotechnol* 2007;2:596–597.

9 Asymmetric Junctionless Transistor

A SRAM Performance Study

Gaurav Saini and Trailokya Nath Sasamal
Department of Electronics and Communication
Engineering, NIT, Kurukshetra, India

CONTENTS

9.1 INTRODUCTION

The revolution in electronic industry started after the first working insulated-gate field effect transistor (FET) by J. Atalla and D. Kahng in 1959 at Bell laboratory. However, the concept of insulated-gate FET had been given by J. E. Lilienfeld in 1926 [1]. The metal-oxide semiconductor (MOS) devices took over the market space since the evolution of CMOS technology [2]. CMOS technology has enabled the scaling of integrated circuits (ICs) as per Moore's law. Further, semiconductor industries started the use of silicon-on-insulator (SOI) substrate for commercial fabrication [3]. SOI technology offers high performance, lower power consumption, and better radiation hardness in comparison to the bulk devices [3]. Increasing demand of low power consumption, high performance, and small chip area of the ICs motivates the researchers to devise novel device structures. To address the scaling issues with bulk devices, many novel device structures were proposed, such as double-gate MOS (DGMOS) transistor [4, 5], triple-gate MOSFET [6], fin field effect transistor (FinFETs) [7–9], and junctionless transistors (JLTs) [10–13]. JLTs are also found to be strong contenders with conventional inversion-mode (IM) devices for low power operation [14, 15].

Nowaday's low-power system-on-chip (SoC) design impose challenges on on-chip memory design because it consumes a major portion of the SoC area [16]. Stability and write ability of SRAM cells degrades with scaling due to lower supply threshold voltage [17]. It is therefore strongly desirable to have strong data stability, wider read/write margin, and lower access time at lower power supply voltages [18, 19]. To address the aforementioned constraints, many studies have been done using device-circuit co-design principles [18, 20–24]. Various asymmetric MOS devices are proposed to enhance the performance of SRAM cell [20–23, 25] and also analog performance [26].

It is evident from the study that a dual-kS JLT device improves the device electrostatic integrity (EI) along with marginal increasing the parasitic capacitances [26]. Therefore, it is envisaged to use the asymmetric junctionless device to address the challenges for possible improvement in the performance of an SRAM cell, especially at lower supply voltages.

9.2   DEVICE STRUCTURE

The device and device-circuit co-design simulation study is performed using Sentaurus TCAD simulator [27]. The device structure of n-channel dual-kS JLT is depicted in Figure 9.1. The gate length (L_g) and silicon channel thickness (T_{si}) are chosen to be 20 nm and 10 nm, respectively. The source/drain (S/D) extension regions (L_{ext}) and gate oxide thickness (T_{ox}) are set to 20 nm and 1.7 nm, respectively. A constant doping concentration of $10^{19}/cm^3$ is considered for source-channel-drain regions. In this work, a comparative study is performed considering three different types of devices named conventional, dual-kS, and dual-kD JLTs. The conventional JLT structure contain low-*k* spacers that cover S/D extension regions completely ($L_{sp,lk} = L_{ext}$). The proposed dual-kS device contains dual-*k* spacers on the source side ($L_{ext} = L_{sp,lk}+L_{sp,hk}$) and a low-*k* spacer only on the drain side ($L_{ext} = L_{sp,lk}$), as given in Figure 9.1a. The dual-kD JLT structure contains dual low-*k* spacers ($L_{ext} = L_{sp,lk}+L_{sp,hk}$) on the drain side and a low-*k* spacer only ($L_{ext} = L_{sp,lk}$) on the source side. The values of $L_{sp,lk}$ and $L_{sp,hk}$ were set to 5 nm and 15 nm, respectively. SiO_2 and TiO_2 are used as

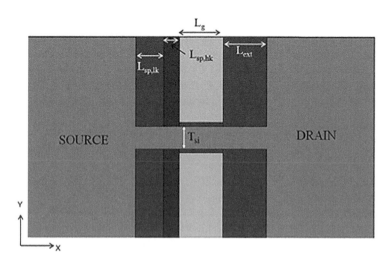

FIGURE 9.1   Device structure of the proposed n-channel dual-kS JLT.

spacer dielectric materials. The threshold voltage is set to 300 mV for both n-type ($V_{tn} \sim 300$ mV) and p-type ($V_{tp} \sim -300$ mV) device structures. Metal gate work function is tuned to get the desired value of threshold voltage for both n-type and p-type devices. Supply voltage (V_{DD}) varied from 0.5 V to 0.8 V. A density gradient quantum correction model is used to perform the simulation study. In addition to it, Auger, SRH recombination/generation, and Lombardi models are considered during simulation [27].

9.3 DEVICE SIMULATION AND ELECTROSTATICS

In this section, the device electrostatics are compared considering the conduction band energy (CBE) profiles of different types of devices. Figure 9.2 shows the CBE diagram of conventional and asymmetric devices (dual-kS and dual-kD). The CBE is plotted along the channel length at $V_{GS} = V_{DS} = 0.0$ V, $V_{GS} = 0.0$ V, and $V_{DS} = 0.8$ V. The CBE diagram shows that the dual-kS JLT structure is the least affected by the drain potential, followed by conventional and dual-kD structures. The value of drain-induced barrier lowering (DIBL) is found to be 15 mV/V for dual-kS JLT, which is superior to conventional (68 mV/V) and dual-kD (87 mV/V) JLT structures. This indicates that the proposed JLT shows improvement in the short channel effects (SCEs) compared to conventional and dual-kD structures. The above-mentioned findings of asymmetric JLT devices can be useful in improving the performance of a SRAM cell, where bidirectional control over the current flow is required during different modes of operation.

Figure 9.3 shows a CBE diagram of conventional, dual-kS, and dual-kD JLTs. The CBE is plotted along the channel length and considering the center point of the

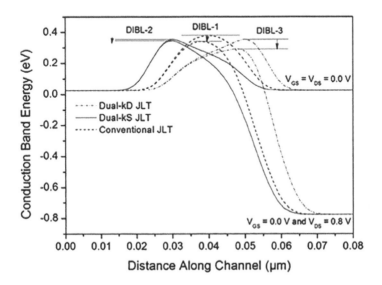

FIGURE 9.2 CBE diagram of different JLTs along the gate length at $V_{GS} = V_{DS} = 0.0$ V, $V_{GS} = 0.0$ V, and $V_{DS} = 0.8$ V.

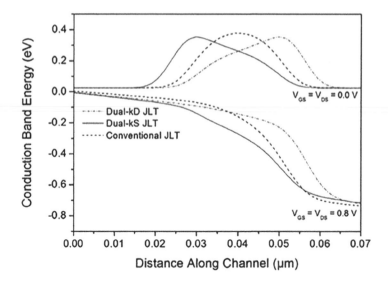

FIGURE 9.3   CBE diagram of different JLTs along the gate length at $V_{GS} = V_{DS} = 0.0$ V and $V_{GS} = V_{DS} = 0.8$ V.

channel at $V_{GS} = V_{DS} = 0.0$ V and $V_{GS} = V_{DS} = 0.8$ V. In the case of dual-kS/dual-kD JLT structures, a shift is observed in the CBE profiles toward the source/drain side, respectively compared with the conventional JLT. Therefore, the proposed asymmetric JLT shows different device electrostatics when source and drain polarities are interchanged. The reduction in energy barrier peak of JLT devices leads to the flow of current from the source to the drain region. The reduction in the energy barrier peak of dual-kS structure is observed to a larger extent, followed by dual-kD and conventional JLTs. Therefore, dual-kS JLT gives highest ON-current (I_{on}), followed by dual-kD and conventional JLTs. The value of I_{on} is found to be 416 µA/µm for the dual-kS, 386 µA/µm for the dual-kD, and 299 µA/µm for the conventional JLT structure.

9.4   SRAM CELL

Figure 9.4 shows the proposed dual-kS JLT–based six-transistor SRAM cell named dual-kS SRAM. The dark thick line shows the dual low-*k* spacer side of asymmetric JLT structure. All asymmetric devices used to design pull-up and pull-down networks act as dual-kS JLT. During read or write operation, source and drain terminals of access transistors interchange depending on the polarity. Therefore, access transistors act as a dual-kS or a dual-kD structure considering the polarity of drain and source terminals.

By assuming that the initial logic at node Q is "high" and QB is "low," the working of the SRAM cell can be explained as follows. During a read operation, the bit-line (BL) and bit-line-bar (BLB) lines are pre-charged to logic "high," and when

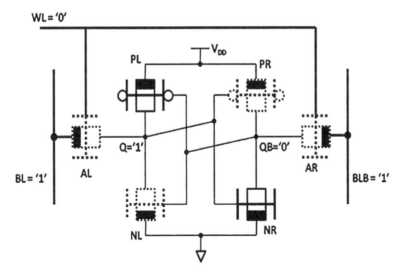

FIGURE 9.4 Proposed dual-kS JLT 6T SRAM cell.

world-line (WL) goes to logic "high," reading of the cell begins. Thereafter, voltage at node QB begins rising toward a certain level called read voltage (V_{read}). The successful read operation requires V_{read} to be lower than the threshold voltage of the other inverter. The read failure can be avoided by keeping the strength of transistor NR greater than that of transistor AR. During the write operation, when WL goes high, node Q starts discharging toward a logic level called write voltage (V_{write}). A successful write can be achieved by keeping V_{write} below the threshold voltage of the other inverter. To improve write stability, the strength of an access transistor (AL) should be higher than that of a pull-up transistor (PL). It is evident from the above discussion that there is a conflict in the strengths of access transistors AL and AR. The main focus of this chapter is to address this conflict by using the asymmetric JLT structure.

9.5 RESULTS AND DISCUSSION

Figure 9.5 shows the hold static noise margin (SNM) of SRAM cells realized using conventional and dual-kS JLTs with respect to supply voltage. Hold SNM of dual-kS JLT SRAM is enhanced by 7.6% and 3.3% when compared to conventional cell at $V_{DD} = 0.5$ V and 0.8 V, respectively. Reduction in supply voltage enhances the percentage improvement in the Hold SNM of the proposed cell when compared to conventional SRAM cell. The read margin of conventional and dual-kS JLT-based 6T SRAM cell is depicted in Figure 9.6. The read margin of the proposed SRAM is improved by 7.4% and 10.1% compared with conventional JLT at $V_{DD} = 0.5$ V and 0.8 V, respectively. The write margin of conventional and dual-kS JLT-based SRAM is depicted in Figure 9.7. The write margin of the presented SRAM is reduced by 9.4% compared with the conventional JLT at $V_{DD} = 0.8$ V. However, no change in the write margin of the presented SRAM cell is observed when compared to conventional JLT at V_{DD} of 0.5 V and 0.6 V.

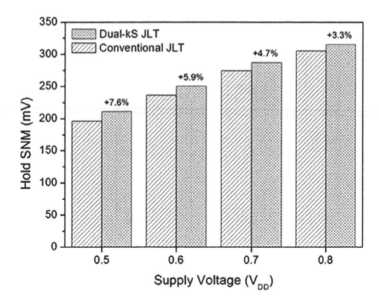

FIGURE 9.5   Hold SNM of conventional and dual-kS JLT-based SRAM cell as a function of V_{DD}.

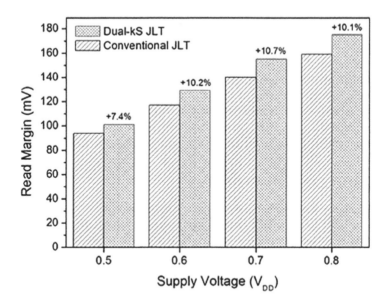

FIGURE 9.6   Read margin of conventional and dual-kS JLT-based SRAM cell as a function of V_{DD}.

The read delay of conventional and dual-kS JLT-based 6T SRAM cell with respect to V_{DD} is shown in Figure 9.8. The read delay of the proposed SRAM cell is found to be improved by 20.6% compared with the conventional JLT at $V_{DD} = 0.8$ V. Interestingly, the percentage delay is improved with the downscaling of V_{DD} from 0.8

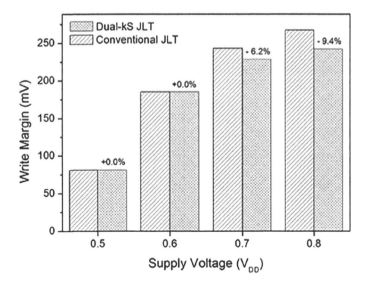

FIGURE 9.7   Write margin of conventional and dual-kS JLT-based SRAM cell as a function of V_{DD}.

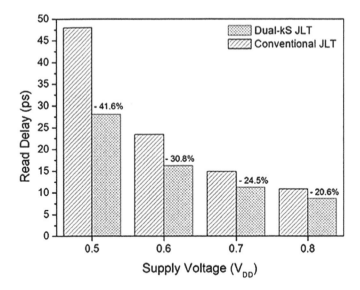

FIGURE 9.8   Read delay of conventional and dual-kS JLT-based SRAM cell as a function of V_{DD}.

V to 0.5 V. The read delay of the proposed SRAM improves by 41.6% compared with conventional structure at $V_{DD} = 0.5$V.

Figure 9.9 shows the write delay of a conventional and dual-kS JLT-based SRAM cell with respect to V_{DD}. The write delay of the proposed SRAM cell is found to be improved by 6.5% compared with a conventional JLT at $V_{DD} = 0.8$V. The percentage improvement in the write delay of the proposed cell is also improved with the

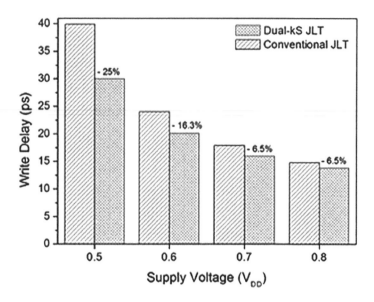

FIGURE 9.9   Write delay of conventional and dual-kS JLT-based SRAM cell as a function of V_{DD}.

TABLE 9.1

Hold, Read, and Write Margins of Conventional and Dual-kS JLT-Based SRAM Cell Structures

V_{DD} (V)	Hold SNM (mV)		Read margin (mV)		Write Margin (mV)	
	Conventional	Dual-kS	Conventional	Dual-kS	Conventional	Dual-kS
0.5	196	211	94	101	81	81
0.6	236	250	117	129	185	185
0.7	274	287	140	155	243	228
0.8	305	315	159	175	267	242

reduction in V_{DD} from 0.8 V to 0.5 V. The write delay of the proposed structure improves by 25% compared with its conventional counterparts at $V_{DD} = 0.5$V.

At lower supply voltage, dual-kS JLS exhibits strong fringe field coupling at the source end in comparison to the conventional structure. This results in the higher improvement in read and write delay of the proposed structure at lower supply voltage.

Table 9.1 shows the comparison of noise margins of dual-kS and conventional cell structures for V_{DD} ranging from 0.5 V to 0.8 V. Table 9.2 shows the comparison of the read/write delay performance of conventional and proposed SRAM structures at different values of V_{DD} ranging from 0.5 V to 0.8 V. It is observed that proposed JLT structure outperforms the conventional structure in terms of delay, especially at lower supply voltages.

9.6   CONCLUSION

The device electrostatics of an asymmetric spacer JLT are explored using extensive device simulation. Effects of source/drain side spacers are studied on the device performance parameters. An asymmetric JLT device (dual-kS JLT) provides improvement in

TABLE 9.2

Delay Performance of Conventional and Dual-kS JLT-Based SRAM Cell Structures

V_{DD} (V)	Read Delay (ps)		Write Delay (ps)	
	Conventional	Dual-kS	Conventional	Dual-kS
0.5	47.96	28.01	39.9	29.93
0.6	23.39	16.19	14.88	11.23
0.7	14.88	11.23	17.89	15.95
0.8	10.86	8.62	14.8	13.84

DIBL and ON-current compared to conventional and dual-kD JLTs. The dual-kS JLT structure is used to design a 6T SRAM cell. The access transistor acts as a dual-kS device, and after reversal of source and drain polarity it acts as a dual-kD device. The above facts are utilized to overcome the challenges associated with the strengths of access and pull-up/pull-down transistors in different modes of SRAM operation. The proposed SRAM cell using asymmetric JLTs shows improvement in hold margin by 7.6% and 3.3%, read margin by 7.4% and 10.1% at supply voltage (V_{DD}) of 0.5V and 0.8V, respectively, compared to the conventional SRAM cell. The write margin of the proposed design shows no change at $V_{DD} = 0.5V$ and a 9.4% reduction at $V_{DD} = 0.8V$ compared with conventional SRAM cell. The proposed SRAM cell also indicates impressive improvement in read delay by 41.6% and 20.6% and write delay by 25% and 6.5% at V_{DD} of 0.5 V and 0.8 V, respectively, compared to the conventional cell. These findings could be useful in designing future memory devices for ultra-low-power applications.

REFERENCES

[1] J. E. Lilienfeld, "Method and apparatus for controlling electric currents," US1745175 A, 28 January 1930.

[2] F. Wanlass and C. Sah, "Nanowatt logic using field-effect metal-oxide semiconductor triodes," in *IEEE Technical Digest of the International Solid State Circuit Conference*, 1963, pp. 32–33.

[3] G. K. Celler and S. Cristoloveanu, "Frontiers of silicon-on-insulator," *Journal of Applied Physics*, vol. 93, pp. 4955–4978, 2003.

[4] T. Sekigawa and Y. Hayashi, "Calculated threshold-voltage characteristics of an XMOS transistor having an additional bottom gate," *Solid-State Electronics*, vol. 27, pp. 827–828, 1984.

[5] D. Hisamoto, T. Kaga, Y. Kawamoto, and E. Takeda, "A fully depleted lean-channel transistor (DELTA)-a novel vertical ultra thin SOI MOSFET," in *Electron Devices Meeting, 1989. IEDM '89. Technical Digest: International*, 1989, pp. 833–836.

[6] M. C. Lemme, T. Mollenhauer, W. Henschel, T. Wahlbrink, M. Baus, O. Winkler, et al., "Subthreshold behavior of triple-gate MOSFETs on SOI material," *Solid-State Electronics*, vol. 48, pp. 529–534, 2004.

[7] D. Hisamoto, L. Wen-Chin, J. Kedzierski, H. Takeuchi, K. Asano, C. Kuo, et al., "FinFET-a self-aligned double-gate MOSFET scalable to 20 nm," *IEEE Transactions on Electron Devices*, vol. 47, pp. 2320–2325, 2000.

[8] Y. Bin, C. Leland, S. Ahmed, W. Haihong, S. Bell, Y. Chih-Yuh, et al., "FinFET scaling to 10 nm gate length," in *Electron Devices Meeting, 2002. IEDM '02. International*, 2002, pp. 251–254.

[9] H. Xuejue, L. Wen-Chin, K. Charles, D. Hisamoto, C. Leland, J. Kedzierski, et al., "*Sub 50-nm FinFET: PMOS*," in *Electron Devices Meeting, 1999. IEDM '99. Technical Digest. International*, 1999, pp. 67–70.

[10] J.-P. Colinge, C.-W. Lee, A. Afzalian, N. D. Akhavan, R. Yan, I. Ferain, et al., "Nanowire transistors without junctions," *Nature Nanotechnology*, vol. 5, pp. 225–229, 2010.

[11] C.-W. Lee, A. Afzalian, N. D. Akhavan, R. Yan, I. Ferain, and J.-P. Colinge, "Junctionless multigate field-effect transistor," *Applied Physics Letters*, vol. 94, p. 053511, 2009.

[12] J. P. Colinge, A. Kranti, R. Yan, C. W. Lee, I. Ferain, R. Yu, et al., "Junctionless nanowire transistor (JNT): Properties and design guidelines," *Solid-State Electronics*, vol. 65–66, pp. 33–37, 2011.

[13] J. P. Colinge, C. W. Lee, A. Afzalian, N. Dehdashti, R. Yan, I. Ferain, et al., "SOI gated resistor: CMOS without junctions," in *SOI Conference, 2009 IEEE International*, 2009, pp. 1–2.

[14] M. S. Parihar and A. Kranti, "Revisiting the doping requirement for low power junctionless MOSFETs," *Semiconductor Science and Technology*, vol. 29, p. 075006, 2014.

[15] M. S. Parihar, D. Ghosh, and A. Kranti, "Ultra low power junctionless MOSFETs for subthreshold logic applications," *IEEE Transactions on Electron Devices*, vol. 60, pp. 1540–1546, 2013.

[16] K. Kang, H. Jeong, Y. Yang, J. Park, K. Kim, and S. O. Jung, "Full-Swing Local Bitline SRAM Architecture Based on the 22-nm FinFET Technology for Low-Voltage Operation," *IEEE Transactions on Very Large Scale Integration (VLSI) Systems*, vol. 24, pp. 1342–1350, 2016.

[17] K. A. Rashmi and G. A. Armstrong, "6-T SRAM cell design with nanoscale double-gate SOI MOSFETs: impact of source/drain engineering and circuit topology," *Semiconductor Science and Technology*, vol. 23, p. 075049, 2008.

[18] S. M. Salahuddin and M. Chan, "Eight-FinFET fully differential SRAM cell with enhanced read and write voltage margins," *IEEE Transactions on Electron Devices*, vol. 62, pp. 2014–2021, 2015.

[19] E. Seevinck, F. J. List, and J. Lohstroh, "Static-noise margin analysis of MOS SRAM cells," *IEEE Journal of Solid-State Circuits*, vol. 22, pp. 748–754, 1987.

[20] A. Goel, S. K. Gupta, and K. Roy, "Asymmetric drain spacer extension (ADSE) FinFETs for low-power and robust SRAMs," *IEEE Transactions on Electron Devices*, vol. 58, pp. 296–308, 2011.

[21] F. Moradi, S. K. Gupta, G. Panagopoulos, D. T. Wisland, H. Mahmoodi, and K. Roy, "Asymmetrically doped FinFETs for low-power robust SRAMs," *IEEE Transactions on Electron Devices*, vol. 58, pp. 4241–4249, 2011.

[22] A. G. Akkala, R. Venkatesan, A. Raghunathan, and K. Roy, "Asymmetric underlapped sub-10-nm n-FinFETs for high-speed and low-leakage 6T SRAMs," *IEEE Transactions on Electron Devices*, vol. 63, pp. 1034–1040, 2016.

[23] P. K. Pal, B. K. Kaushik, and S. Dasgupta, "High-performance and robust SRAM cell based on asymmetric dual-K spacer FinFETs," *IEEE Transactions on Electron Devices*, vol. 60, pp. 3371–3377, 2013.

[24] A. Bansal, S. Mukhopadhyay, and K. Roy, "Device-optimization technique for robust and low-power FinFET SRAM design in nanoscale era," *IEEE Transactions on Electron Devices*, vol. 54, pp. 1409–1419, 2007.

[25] G. Saini and S. Choudhary, "Improving the performance of SRAMs using asymmetric junctionless accumulation mode (JAM) FinFETs," *Microelectronics Journal*, vol. 58, pp. 1–8, 2016.

[26] G. Saini and S. Choudhary, "Asymmetric dual-k spacer trigate FinFET for enhanced analog/RF performance," *Journal of Computational Electronics*, vol. 15, pp. 84–93, 2016.

[27] S. Inc. (2014). *Synopsys Sentaurus Design Suite*. Available: http://www.synopsys.com

10 Performability Analysis of High-*k* Dielectric– Based Advanced MOSFET in Lower Technology Nodes

Manoj Angara
Koneru Lakshmaiah Education Foundation,
Vaddeswaram, India

Biswajit Jena and Ayodeji Olalekan Salau
Afe Babalola University, Ado-Ekiti, Nigeia

CONTENTS

10.1 INTRODUCTION

The numerous applications and recent developments in semiconductor electronics have led to continuous research and interest in nanoscale devices that are being improved by using different engineering techniques [1, 2]. Recently, nano-scale devices have emerged as new avenues for the designers as well as manufacturers to overcome the scaling issues [3, 4]. The scaling of semiconductor devices has been an issue in the microelectronics industry because of faster device operation with almost the same cost. Although the basic structure of the metal oxide semiconductor (MOS) devices remains the same, the gate length reduces and enters into the sub-nanometer regions. Therefore, introducing scaling techniques to the gate is not the ultimate solution to improve the device performance for improved packing density and low cost [5–9]. So there should be a trade-off between scaling of device dimension and power supply voltage, otherwise the internal high electric field will damage the device irreversibly. This unwanted electric field causes some malfunctioning behavior of the device performance, resulting in short channel effects (SCEs) [10–13]. The persistent scaling of the CMOS technology in the sub-nanometer region alters the physical properties of the device and thus affects the current flow characteristics. Continuous miniaturization of device size with reduced channel length creates some unwanted effects such as channel length modulation, threshold voltage roll-off, punch-through effect, parasitic capacitance increments, and drain-induced barrier lowering (DIBL) [14–17]. In addition, some analog and RF performances of the device also get affected by scaling the device channel length. The objective of scaling the device is to shrink transistor dimension so that the number of transistors can be arranged on a chip. Basically, the scaling factor '*k*' is $\sqrt{2}$, so that the area is reduced to one-half and the number per area increases by a factor of two [18,19]. Integration of more transistors on a single chip is possible due to the possibility of designing smaller components on silicon using optical lithography. As optical lithography enters into the sub-wavelength regime, light diffraction and interference from sub-wavelength pattern feature causes image disorder. Therefore, patterning becomes very difficult without introducing resolution enhancement techniques. As the traditional planar MOSFET design slowed down due to scaling problems, new materials or gate engineering are necessary in order to match the current requirement of semiconductor technologies [20–22]. Evolution of the MOSFET technology has achieved milestones in both device design and performance. Gate engineering technologies changed traditional MOSFET to a MOSFET design with numerous gates, as shown in Figure 10.1. In case of a fully depleted double-gate (DG) and cylindrical surrounding-gate MOSFET, the short channel effects are governed by the electrostatic potential, and in these structures the channel is confined by the gate metal [23–25]. So, gates provide a superior controllability to reduce these types of effects. In order to accommodate a higher number of transistors, the size of the individual transistors

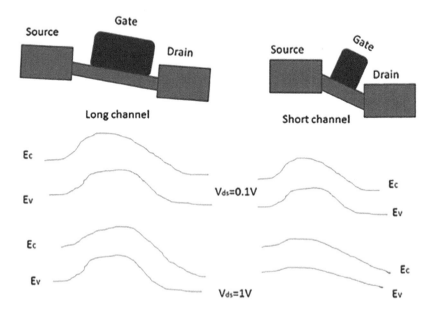

FIGURE 10.1   Long channel vs short channel devices.

shrinks rapidly. Shrinking of transistor size leads to reduced oxide thickness and channel length [26–29]. At the same time effective gate length also reduces and gate controllability over the channel reduces drastically. This brings about uneven and unpredictable behavior of the device performance, which affects the overall performance of the system. As a result of shortening of the device size, several short channel effects (SCEs) such as threshold voltage roll-off, drain induced barrier lowering (DIBL), and channel length modulation (CLM) are caused [30]. At the same time, reduced gate oxide thickness leads to severe quantum mechanical effects and hence results in more leakage current (I_{off}). Large leakage current affects the CMOS characteristics significantly by consuming more power. There are several factors that create short channel effects such as the depletion width of the channel and source/drain depth at the junction. These factors also are scaled down along with channel length and oxide thickness. In order to scale down the channel depletion width, the doping concentration of the channel needs to be increased, as a result of which the OFF-current (leakage current) reduces and controls the threshold voltage. Some pros and cons are generated with higher channel doping. These include reduced carrier mobility in the channel, as an increase in the vertical electric field creates more impurity scattering. These cons should be addressed carefully in order to design a high-performance CMOS device for the next-generation semiconductor industry.

As the semiconductor industry is working on the present technology to match the current requirement, the sizes of semiconductor devices are getting smaller and smaller. Today, the industry has reached the nm technology node by overcoming the μm technology. As the device size becomes smaller, the drain and source come closer to each other. The electrostatic control of the gate over the channel will be affected by drain voltage as the channel length become smaller. This can be well understood from the energy band diagram analysis shown in Figure 10.1. From the energy band diagram for conduction band and valence band at low drain bias (0.1 V), the energy barrier exists for both the devices. However, for a higher drain bias (1.0 V), band bending is observed with a sharp slope in short channel devices due to interference of drain voltage on gate voltage as shown in the figure. The semiconductor industry requires interlayer dielectric for a variety of technological developments that take place to bring the devices into the next level of technology. Dielectric layers have insulation as well as capacitor used as memories. Silicon dioxide (SiO_2) is considered the best choice to carry out these operations for a long time because it has a large band gap of ~9 eV and high resistance around ~10^{18} Ω cm, which is suitable for dielectric materials [31,32]. Downscaling the thickness of SiO_2 to 2 nm was reported in 2001, and the physical limit was also defined where the band gap will not keep its original value anymore. Although the dielectric constant of SiO_2 can be tuned by adding some nitrogen component to it to extend k = 6, this is not enough and is a complex process. This gives rise to very low leakage current with well-defined interface with silicon. However, downscaling the devices into lower technology nodes is rejected, as it has some physical limitation below 1 nm. Reducing the physical size of SiO_2 achieves the goal of improved capacitance, but the thin film power consumption increases drastically as tunneling comes into the picture. In order to overcome these issues, high-*k* materials are required to increase the capacitance by reducing the leakage current, as it can be scaled down to a larger extend. A large range of high-*k*

materials are being proposed to mitigate this scaling issue. Another issue that is dis-
covered while selecting high-*k* materials is the deposition of films over the silicon.
However, few techniques are well accepted and proved efficient to carry out this
process. Chemical vapor deposition (CVD) and physical vapor deposition (PVD) are
the two common processes used for deposition. In order to control the thickness of
the deposition, atomic layer deposition (ALD) is commonly used [33,34]. The benefits of replacing high-*k* in place of low-*k* can be well understood by observing the
Equation 10.1, which gives the EOT thickness calculation equation [35,36]. The
equation is given as follows:

$$t_{eq} = t_{high-K} \frac{\in_{low-K}}{\in_{high-K}} = t_{high-K} \frac{3.9}{\in_{high-K}} \tag{10.1}$$

Equation 10.1 is used to calculate the equivalent thickness of the dielectric, when
a high-*k* is used in place of low-*k* dielectric. The use of high-*k* dielectric for scaling
purpose can be easily understood by observing Figure 10.2. In Figure 10.2, the SiO_2
is replaced with HfO_2, which has a dielectric constant of 22. The equivalent oxide
thickness of 1 nm SiO_2 will be ~5 nm HfO_2, which plays a vital role for downscaling
the device into the next technology node.

To solve the scaling challenge, we propose a 3D structure with a 3D view shown
in Figure 10.3 with high-*k* as dielectric and the cut section is shown is 3(b) with all
the regions marked along the y-axis. In order to identify the regions more accurately,
the cut section along the z-axis is shown in Figure 10.3c.

The presence of small-thickness interfacial layer (Si_3N_4) is used for smooth fabrication, as it is difficult to fabricate HfO_2 directly on the silicon surface. Figure 10.4
shows the cross-sectional view of the proposed structure with different dielectrics.
The thickness of the dielectric is kept constant, 2 nm, for all the cases to identify its
superiority more accurately. The proposed structure with different dielectrics is
shown in Figures 10.4a–c. For HfO_2, an interfacial dielectric of thickness 0.5 nm is
used to give a strong bonding and fabrication compatibility with silicon.

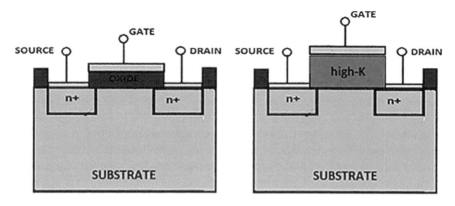

FIGURE 10.2 Equivalent scaling for high-*k* in place of oxide for further scaling in conventional MOSFET.

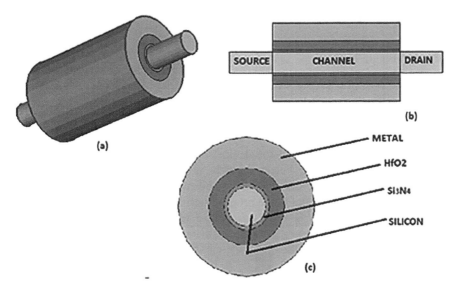

FIGURE 10.3 (a) 3D view of the proposed model, (b) cross-sectional view along the length, and (c) z-cut to identify the individual layers.

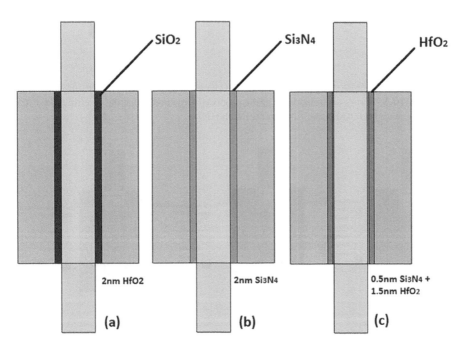

FIGURE 10.4 Cross-sectional view of the proposed structure with (a) SiO_2, (b) Si_3N_4, (c) Si_3N_4, and HfO_2.

10.2 RESULTS AND DISCUSSIONS

The electric field distribution for the three structures with different dielectrics is shown in Figure 10.5.

The value of absolute electric field at a particular region can be identified by comparing the values in the legend. The electric field distribution is taken at a low bias of 0.2 V. Similarly, the electrostatic potential is shown in Figure 10.6.

The transfer characteristics of the three structures is shown in Figure 10.7. Figure 10.7 shows both the linear as well as the logarithmic scale to identify the ON-current and OFF-current of the device. In the linear scale, the superiority of high-*k* dielectric over low-*k* dielectric can be observed clearly. Introducing high-*k* dielectric improves not only the scaling possibility but also the high drain current.

The transfer characteristics of the proposed structure with low-*k* (SiO_2) dielectric is shown in Figure 10.8. In order to analyze the device performance in lower

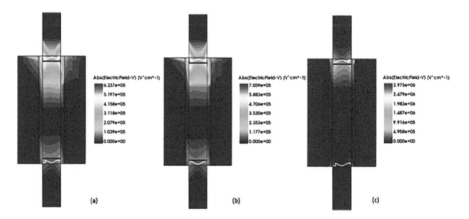

FIGURE 10.5 Cross-sectional view of the proposed structure indicating electric field distribution (a) with SiO_2, (b) with Si_3N_4, and (c) with HfO_2.

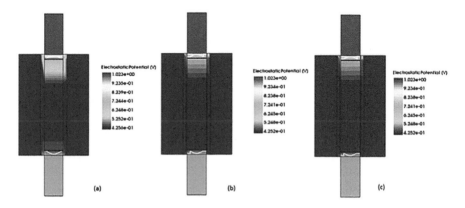

FIGURE 10.6 Cross-sectional view of the proposed structure indicating electrostatic potential distribution (a) with SiO_2, (b) with Si_3N_4, and (c) with HfO_2.

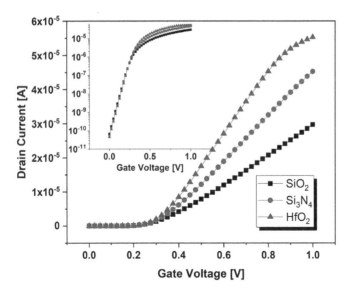

FIGURE 10.7 Comparison of transfer characteristics of the structure of three different dielectrics both in linear and logarithmic scales.

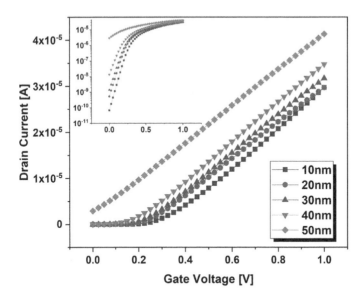

FIGURE 10.8 Comparison of transfer characteristics of the structure with SiO_2 as dielectric in both linear and logarithmic scales for different gate lengths.

technology node, the gate length is taken from 50 nm to 10 nm. From Figure 10.8 it is observed that the logarithmic form of the transfer function indicates an increase in the drain current as the gate length is reduced, but at the same time the OFF-current increases, which eventually degrades the device performance.

The transfer characteristics of the proposed structure with high-*k* dielectric as Si_3N_4 and HfO_2 are shown in Figures 10.9 and 10.10, respectively.

In order to determine the devices performance in low as well as high technology nodes, the gate length is scaled down from 50 nm to 10 nm, and the corresponding results are documented. From the figures, the superiority of high-*k* dielectric over low-*k* dielectric can be determined for both the cases.

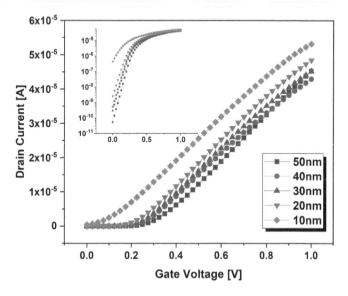

FIGURE 10.9   Comparison of transfer characteristics of the structure with Si_3N_4 as dielectric in both linear and logarithmic scales for different gate lengths.

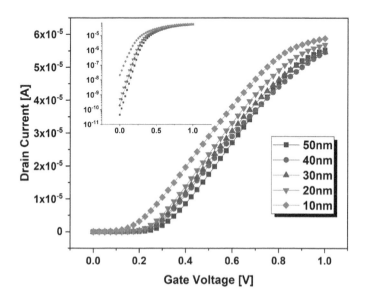

FIGURE 10.10   Comparison of transfer characteristics of the structure with HfO_2 as dielectric in both linear and logarithmic scales for different gate lengths.

10.3 CONCLUSION

High-*k* dielectric is emerging as an important component to scale down MOSFET devices into lower technology nodes. Many high-*k* materials are currently under investigation to replace SiO_2 because of its performance and cost. With high expectations of standard and performance from the semiconductor industry, the research efforts to replace SiO_2 as dielectric are extremely challenging and still undergoing trials. However, Si_3N_4 and HfO_2 are considered as the best alternatives as dielectric for miniaturization of the devices. While some hopeful results have been obtained in this work, the search for the ideal high-*k* material is continuous and leads toward various kinds of composite high-*k* materials and even novel device structures.

REFERENCES

[1] Y. Taur and T. H. Ning, *Fundamentals of modern VLSI devices*. Cambridge: Cambridge University Press, 2013.

[2] S. Tayal and A. Nandi, Enhancing the delay performance of junctionless silicon nanotube based 6T SRAM, *Micro & Nano Letters*, vol. 13, pp. 965–968, 2018.

[3] J. P. Colinge, *Silicon-on-insulator technology: Materials to VLSI*. New York: Springer Science and Business Media, 2004.

[4] W. Y. Choi and H. K. Lee, Demonstration of hetero-gate-dielectric tunneling field-effect transistors (HG TFETs), *Nano Convergence*, vol. 3, p. 13, 2016.

[5] B. Jena, S. Dash, K. P. Pradhan, S. K. Mohapatra, P. K. Sahu, and G. P. Mishra, Performance analysis of undoped cylindrical gate all around (GAA) MOSFET at sub threshold regime, *Advances in Natural Sciences: Nanoscience and Nanotechnology*, vol. 6, pp. 035010, 2015.

[6] V. P. Georgiev et al., Experimental and simulation study of silicon nanowire transistors using heavily doped channels, *IEEE Transactions on Nanotechnology*, vol. 16, no. 5, pp. 727–735, 2017.

[7] S. Tayal and A. Nandi, Optimization of gate-stack in junctionless Si-nanotube FET for analog/RF performance, *Materials Science in Semiconductor Processing*, vol. 80, pp. 63–67, 2018.

[8] M. K. Anvarifard and A. A. Orouji, Improvement of self-heating effect in a novel nanoscale SOI MOSFET with undoped region: A comprehensive investigation on DC and AC operations, *Super Lattices and Microstructures*, vol. 60, pp. 561–579, 2013.

[9] J. Madan and R. Chaujar, Interfacial charge analysis of heterogeneous gate dielectric-gate all around-tunnel FET for improved device reliability, *IEEE Transactions on Device and Materials Reliability*, vol. 16, pp. 227–234, 2016.

[10] S. Tayal, S. Gupta, A. Nandi, A. Gupta, and S. Jadav, Study of Inner-gate Engineering Effect on Analog/RF Performance of Conventional Si-Nanotube FET, *Journal of Nanoelectronics and Optoelectronics*, vol. 14, pp. 953–957, 2019.

[11] T. Busani and R. A. B. Devine, The importance of network structure in high-k dielectrics: $LaAlO_3$, Pr_2O_3, and Ta_2O_5, *Journal of Applied Physics*, vol. 98, no. 4, p. 044102, 2005.

[12] D. P. Brunco, A. Dimoulas, N. Boukos et al., Materials and electrical characterization of molecular beam deposited CeO_2 and CeO_2/HfO_2 bilayers on germanium, *Journal of Applied Physics*, vol. 102, no. 2, p. 024104, 2007.

[13] S. Tayal and A. Nandi, Effect of high-K gate dielectric in-conjunction with channel parameters on the performance of FinFET based 6T SRAM, *Journal of Nanoelectronics and Optoelectronics*, vol. 13, pp. 768–774, 2018.

[14] P. Keerthana, P. P. Babu, T. A. Babu, and B. Jena, Performance analysis of GAA MOSFET for lower technology nodes, *Journal of Engineering Science and Technology Review*, vol. 13, pp. 39–43, 2020.

[15] J. T. Park, J. Colinge, and C. Diaz, Pi-gate SOI MOSFET, *IEEE Electron Device Letters*, vol. 22, no. 8, pp. 405–406, 2001.

[16] N. H. E. Weste and D. M. Harris, *CMOS VLSI design*, 4th edition, Addison-Wesley, 2011.

[17] S. M. Kang and Y. Leblebici, *CMOS digital integrated circuits*, 3rd edition, McGraw-Hill, 2017.

[18] K. Nayak, et al., Negative differential conductivity and carrier heating in gate-all-around Si Nanowire FETs and its impact on CMOS logic circuits, *Japanese Journal of Applied Physics*, vol. 53, pp. 4–16, 2014.

[19] J. J. Gu, Y. Q. Liu, Y. Q. Wu, R. Colby, R. G. Gordon, and P. D. Ye, First experimental demonstration of gate-all-around III–VMOSFETs by top-down approach, *IEEE International Electron Devices Meeting*, vol. 11, pp. 769–773, 2011.

[20] N. Singh et al., High-performance fully depleted silicon nanowire (diameter ≤ 5 nm) gate-all-around CMOS devices, *IEEE Electron Device Letters*, vol. 27, no. 5, pp. 383–386, 2006.

[21] Q. Zhang et al., Novel GAA Si Nanowire P-MOSFETs with excellent short channel effect immunity via an advanced forming process, *IEEE Electron Device Letters*, vol. 65, pp. 464–467, 2018.

[22] J. T. Park, J. Colinge, and C. Diaz, Pi-gate SOI MOSFET, *IEEE Electron Device Letters*, vol. 22, no. 8, pp. 405–406, 2001.

[23] M. D. Marchi, D. Sacchetto, S. Frache, J. Zhang, P. Gaillardon, Y. Leblebici, and G. D. Micheli, Polarity control in double-gate, gate-all-around vertically stacked silicon nanowire FET, *Proceedings of the International Electron Devices Meeting*, San Francisco, CA, pp. 4–8, 2012.

[24] B. Jena, S. Dash, and G. P. Mishra, Improved switching speed of a CMOS inverter using work-function modulation engineering, *IEEE Transaction on Electron Devices*, vol. 65, pp. 2422–2425, 2018.

[25] D. Nagy, G. Indalecio, A. J. García-Loureiro, M. A. Elmessary, K. Kalna, and N. Seoane, FinFET versus gate-all-around nanowire FET: Performance, scaling, and variability, *IEEE Journal of the Electron Devices Society*, vol. 6, pp. 332–340, 2018.

[26] S. Tayal, V. Mittal, S. Jadav, S. Gupta, A. Nandi, and B. Krishan, Temperature sensitivity analysis of inner-gate engineered JL-SiNT-FET: An analog/RF prospective, *Cryogenics*, vol. 108, pp. 103087, 2020.

[27] S. Tayal, P. Samrat, V. Keerthi, B. Vandana, and S. Gupta, Channel thickness dependency of high-K gate dielectric based double-gate CMOS inverter, *International Journal of Nano Dimension*, vol. 11, pp. 215–221, 2020.

[28] N. H. E. Weste and D. M. Harris, *CMOS VLSI Design*, 4th edition, Addison-Wesley, 2011.

[29] Y. Sugiyama, S. Pidin, and Y. Morisaki, Approaches to using Al_2O_3 and HfO_2 as gate dielectrics for CMOSFETs, *Fujitsu Scientific & Technical Journal*, vol. 39, pp. 94–105, 2003.

[30] S. Guha and V. Narayanan, High-κ/Metal gate science and technology, *Annual Review of Materials Research*, vol. 39, pp. 181–202, 2009.

[31] S. T. Wara, A. Abayomi-Alli, and A. O. Salau, Evaluation of the dielectric properties of local cereals in the audio frequency range, *5th IEEE International Conference on Signal Processing, Computing and Control (ISPCC)*, pp. 313–318, 2019. doi:10.1109/ispcc48220.2019.8988329

[32] J. C. Lee, Single-layer thin HfO2 gate dielectric with n+ polysilicon, Proceedings of IEEE Symposium on VLSI Technology, NJ, pp. 44–45, 2000.

[33] H. S. Baik and S. J. Pennycook, Interface structure and non-stoichiometry in HfO$_2$ dielectrics, *IEEE Applied Physics Letter*, vol. 85, pp. 672–674, 2009.

[34] A. P. Huang, Z. C. Yang, and P. K. Chu, Hafnium-based high k gate dielectrics, *Proceedings of Advances in Solid State Circuits Technologies*, pp. 333–350, 2010.

[35] M. Fadel and O. Azim, A study of some optical properties of hafnium dioxide thin films and their applications, *Journal of Applied Physics Materials Science and Processing*, vol. 66, no. 3, pp. 335–343, 1997.

[36] B. R. Dorvel et al., Silicon nanowires with high-k hafnium oxide dielectrics for sensitive detection of small nucleic acid oligomers, *ACS Nano*, vol. 6, no. 7, pp. 6150–6164, 2012.

Index

Page numbers in *italics* refer to figure; Page numbers in **bold** refer to table.